Perspectives on the History of Chemistry

Series Editor

Seth C. Rasmussen, Department of Chemistry and Biochemistry, North Dakota State University, Fargo, ND, USA

Commonly described as the "central science", chemistry and the chemical arts have an extremely long history that is deeply intertwined with a wide variety of other historical subjects. Perspectives on the History of Chemistry is a book series that presents historical subjects covering all aspects of chemistry, alchemy, and chemical technology.

Potential topics might include:

- An updated account or review of an important historical topic of broad interest
- Biographies of prominent scientists, alchemists, or chemical practitioners
- Translations and/or analysis of foundational works in the development of chemical thought

The series aims to provide volumes that advance the historical knowledge of chemistry and its practice, while also remaining accessible to both scientists and formal historians of science. Volumes should thus be of broad interest to the greater chemical community, while still retaining a high level of historical scholarship. All titles should be presented with the aim of reaching a wide audience consisting of scientists, chemists, chemist-historians, and science historians.

All titles in the book series will be peer reviewed. Titles will be published as both printed books and as eBooks. Both solicited and unsolicited manuscripts are considered for publication in this series.

E. Allen Driggers

Early Nineteenth Century Chemistry and the Analysis of Urinary Stones

Springer

E. Allen Driggers
Tennessee Technological University
Cookeville, TN, USA

ISSN 2662-4591 ISSN 2662-4605 (electronic)
Perspectives on the History of Chemistry
ISBN 978-3-031-34972-0 ISBN 978-3-031-34973-7 (eBook)
https://doi.org/10.1007/978-3-031-34973-7

This Springer imprint is published by the registered company Springer Nature Switzerland AG
The registered company address is: Gewerbestrasse 11, 6330 Cham, Switzerland

This book is lovingly dedicated to Dr. Laura Elizabeth Smith

Preface

This book was my divorce book. This book was my pandemic book. This book was my colectomy book. This was the book I wrote in pandemic exile. This book was also an attempt to make sense of my own urinary stone experience. The revising of this book occurred during a moment when my life changed to be better than I could have ever imagined, compared to when first starting to write it. This book was written in several places and in several life situations. The book was written during an academic fellowship, in four different cities, and in two distinct phases in my life, and two professional ranks. It was written with a job, without a job, and then a job again. Though a lot of people have come and gone, those that stayed around to see its completion, this is solely your book, as you were a ride or die type, and never have, and never will give up. You are truly the best of people. And for those people who are not good: I did this despite you.

The essence of this project started when I suffered from my own urinary stone. I was a young enthusiastic graduate student at the University of South Carolina. The heat and humidity of Columbia, South Carolina, combined with the lack of hydration, caused me to produce the symptoms of a stone, that seemed most mysterious to myself, especially the time I started producing purple and blood colored urine. This was quite a surprise! A urologist quickly passed a comforting diagnosis, but emphasized that I did have to pass it myself, which was intimidating. I passed the stone prior to assisting in a lecture and walked away with the passed stone for further chemical analysis. After reading, accidently, several historical accounts of practitioners in the eighteenth century trying to treat stones, I felt a historical resonance. I quickly read over the work of Joseph Ioor Waring and his *History of Medicine in South Carolina* and found a urologist by the name of Edward Darrell Smith. Smith's small but dense case histories gave me a window into urological practice in South Carolina. After discussing the project with my advisor, Dr. Ann Johnson, she did not hesitate to encourage me to write a dissertation on urological history. She, and later I, saw the potential in the project as it was potentially bigger than simply a history of urology, but a project that could involve the history of humoral theory, the history of chemistry, and examine the transition in medicine away from humoral medicine. I was hooked and kept reading, and that enthusiasm allowed me to travel to the cities of London, Charleston, and Philadelphia. Continuing to rework, revise, and rewrite this project proved challenging, as many presses saw the project as narrow and did not want to take a chance on the project. I continued to revise the now book project through personal challenges and unexpected sabotage.

However, while experiencing a run of bad luck with publications during the project, I met Laura Elizabeth Smith. She burst forth from her aunt's car after a research trip to Washington DC after just finishing her comprehensive exams and I felt the earth move, and it was been moving ever since. She was been encouraging about the project; even after she watched me receive more rejections than I care to remember. She never let me give up on the project and has both supported and encouraged me in all endeavors. We supported each other through an eight hour distance, and then one gigantic pandemic sized move.

This book continued to live and change, and it was revised during the lives of two women who grew to become amazing people. Mildred Penelope Driggers met Laura when I was reforming my ideas and book after an unfortunate sabotage. Laura and Millie met each other almost five years ago, and they have both become really interesting people. Laura Elizabeth Smith grew into the teacher and research she always wanted to be, completing her Ph.D. at the University of Arkansas and teaching many classes. She was already a mom to dogs, but she gained both another dog, little Homer, who is currently two, and Millie who was two at the time. I often thought I cried tears of sadness as time passed as dogs and children get older (and sometimes cats) but I think they are tears of joy at how amazing seeing your spouse grow and change, and accomplish things that they didn't realize they could achieve, and I realize that life is even better than I could have ever realized after a unforeseen series of events came to my front door in 2017.

Every day that Laura comes in my office it is daily amazing realization that she became everything she aspired to (and even more) when Millie and I met her for ice cream and I am able to watch her do new and more incredible things every day. I used to think one of the happiest days of my life would be winning the Francis Parkman prize or something silly like that, but the happiest days of my life is every Tuesday when Millie lays on Laura after she gets back from dance and tells her about all of the things going on in her life and asks her all of the questions she asks me that I often answer with "ask Laura about that." Laura listens patiently as Millie plans the person she wants to become and aspire to be, and Laura offers her nothing but unconditional love and support. Laura performs the magic that allows Millie to sleep and then Laura tells me about her classes and the research and crafts she aspires to work on, and I am happy at the possibilities of seeing those new things on our Christmas tree or our house, at church, and in our lives. My heartfelt thanks go to her for being an incredible mother, raising our daughter Mildred Penelope, to be a confident and contributing member of society. This book, dedicated to Dr. Laura Elizabeth Smith, serves as a memorial that with the right magic lead can be turned into gold.

Cookeville, USA E. Allen Driggers

Acknowledgements

Thanks goes to my colleagues at Tennessee Technological University in Cookeville, Tennessee. We have a fantastic department. Thank you, Amy Foster; you do so many jobs to help all of us. Thank you for all the assistance, help, car rides, and free counseling I have received over the years. I'd like to thank my mentor here at Tech, Dr. Kent Dollar. He shepherded me through tenure, promotion, and learning about the academy. Dr. Dollar was kind enough to lend his time and efforts to support me professionally. I appreciate him serving as chair and helping me receive several grants to support this book. Thank you, Dean Jeff Roberts, for assistance with the project! Thank you, Susan Laningham, for all the childcare and dance lessons for Millie while I was writing this book. Thank you to Ted Pelton for getting me into old timey baseball. Laura and I were also very lucky to have such great dinner dates with Phillip Davis and Shirley Laird. We have been befriended by the Lord and Lady of Sparta, Troy Smith and Robin Smith, and we so very much appreciate their friendship and neighborly presence.

The writing of this book benefited from generous financial support from a number of sources. Beth Lander was an amazing source of information, a kind friend and librarian, and brought this project to life. I was incredibly lucky to get to visit the Mutter through several Wood Institute Travel Grants. Earlier in my career, I received the Edelstein Dissertation Writing Fellowship at the Chemical Heritage Foundation (now Science History Institute) for the 2014–2015 academic year in Philadelphia, Pennsylvania. My travel to London was supported by the Provost Office's SPARC Grant at the University of South Carolina. The Chemical Heritage Foundation (now Science History Institute) also funded my archival research for two weeks at their Philadelphia-based archives. I also received two years of funding from the Tennessee Technological University Faculty Research Committee. I want to give a special acknowledgement to the kind gifts of Mike Winchester to Tennessee Technological University's History Department. Special thanks for the mentoring I received at the Science History Institute by Dr. Carin Berkowitz, Dr. Ronald Brashear, and library help from Ashley Augustyniak. Also thank you for encouraging words by Dr. Laura Cruz, who encouraged me to keep trying to find a publishing home for the project.

Many people suffered through reading many parts of this book. Thank you, Dr. David Dangerfield, for your insightful comments on this book in its various forms and your continued friendship; you are a fantastic scholar and father, and the true Dean of South Carolina history. Through your thoughts and prayers, I have been lifted higher than I thought I could ever go when I was a silly graduate student at University of South Carolina. I appreciate you and your family and want to say that you are a positive testament of South Carolina, and an artist chef.

I'd also wish to thank Dr. Larry Glickman at Cornell who kindly kept me afloat and my spirits high, with Ann Johnson, while I underwent major surgery and during the death of my father while I was attending the University of South Carolina. There were some truly kind spirits at USC. I'd like to thank my surgeon in Columbia, Dr. Ronald "Trey" Craft, III for some truly kind cuts during my partial colectomy. It amazes me daily how much healing can be done with a scalpel.

Two people who had to exit the project for bigger and better things were my doctoral advisor and my father. Both succumbed to the scourge of cancer too early, and I think about them every day. Dr. Ann Johnson, of Cornell and the University of South Carolina, held court in two worlds, history and philosophy, but was humble in all realms. Ann was called a professor's professor in so many chronicles of her life and that is the best way to describe her. She was open to this project about urinary stones, mentored me about how to complete it, and sloughed through many drafts of this project. Every kindness I show a student I do in her memory.

My father gave me confidence to continue in history, whether he realized it or not. Though not the scholar himself, he helped me feel comfortable in the academy. I got a summer job at the School of Design at North Carolina State and found the environment stimulating and exuberant. I was lucky enough to get a scholarship for the children of staff at North Carolina State and pursued my undergraduate degree there. One weekend day at Bob Evans Restaurant, I was pessimistic about the value of history, and he reminded me of the value of history. I was able to renew the value in what I was doing, and he supported me through my Ph.D. at the University of South Carolina. I miss our conversations every day.

My mother has stimulated my life trajectory into research. She explained the world of academia early on in my life and took me to enriching concerts, lectures, and art shows. She always encouraged my intellectual pursuits, and I thank her for her support. My mom has done a lot of encouraging and supporting of this project. She is also an excellent grandma! Laura and I have been grateful for her watchful eye on Millie and refreshing visits to Durham. I'd also like to thank the Smiths, both Sarah and Grant, whom I recently gained as family in 2019. They both have been helpful and encouraging of the project, and importantly provided much needed childcare.

My undergraduate mentor, Dr. Larry Blanton, also spurred my intellectual curiosity and engagement at North Carolina State, and inducted me into the Zeta chapter of Phi Beta Kappa. Through mentorship, individual tutorials, amazing house concerts, introductions to Lubbock Texas, and summer research grants, he helped mold and support the scholar that I am today. As a professor, I aspire and encourage my own undergraduate students in the same helpful and kind way as Dr. Blanton.

Amy Hill at Tennessee Tech breathed fresh life into this project. Without her invaluable comments and labor into helping me revise this book, I might have been editing it forever, and it might still be sitting on the shelf. Thank you for your tireless work on this project! I sincerely appreciate your selfless gift of your time.

I would also like to thank many presses that did not limit this book, as they did not have the vision that Springer had for the project. I would like to thank Dr. Seth Rasmussen for taking a chance on a first-time author and providing kind editing and advice about how to improve the project; I am extremely lucky to be in this exciting series! I would also like to extended my thanks to the Pennsylvania State Press for permission to utilize the articles Driggers E (2019) The Chemistry of Blackness: Benjamin Rush, Thomas Jefferson, Everard Home, and the Project of Defining Blackness Through Chemical Explanations. Crit Phil Race 7:372–391, and I would also like to thank Sage for permissions for Driggers E (2020) "A Clue in a Labyrinth:" The Medico-Chemistry of the Stone in the Life of Edward Darrell Smith. J Med Bio 28:232–245.

The images in this book were kindly provided by the Science History Institution (formerly the Chemical Heritage Foundation) and the Reynolds-Finley Historical Library, the University of Alabama at Birmingham. I would like to thank Annabel Pinkney of Science History Institute and the Othmer Library. During my time as the Reynolds-Finley Fellow, I was assisted by Anna Kaetz and Peggy Balch. My time at the Reynolds-Finley Historical Library was truly fantastic, and I appreciate Peggy and Anna's friendship!

My wife, Laura, has shown me a world that I could not have thought possible five years ago. From ballet in the kitchen, to helping a student through a medical emergency in Rome, she has been a companion's companion, and a teacher's teacher. Also, she's been my lockdown spouse, as we spent our first months of marriage in exile in Arkansas. The adventure continues as she is my office neighbor. I was able to witness so many amazing things in our life together such as awaiting the defense of her own dissertation, which she dealt with mightily, as a hero would slay a dragon. She loves me with a heart that has no limit. If the experiences of life were a pizzeria, her response at a Mellow Mushroom in Chattanooga would best embody my experiences with her: "We are gonna tear this place up." I am looking forward to a future where we grow together as one: Genesis 2:24, and I strive for Ephesians 4:2. Though we will never know what the future holds, but "I swear it's so hard to bear it/And I'd never make it through without you around/No I'd never make it through without you around." I was editing this book while Dr. Laura Elizabeth Smith defended her dissertation on November 18, 2022, at high noon. She defended her dissertation while I anxiously worked on my revisions in the hallway outside; she passed and this manuscript's dedication had to be updated to fit the new title.

June 2023 E. Allen Driggers

Contents

About the Author

E. Allen Driggers is an associate professor of the history of science at Tennessee Technological University in Cookeville, Tennessee, United States of America. He teaches courses in the history of science, British History, food studies, engineering studies, and the history of South Asia. He has several publications in early nineteenth century medicine and science, from the history of botany, engineering, dinosaurs, and urinary stones. He is working on his next book project about the history of medicine in British India. He is married to fellow academic Dr. Laura Elizabeth Smith, whom he sometimes collaborates with on articles regarding science and race. They have two dogs, one cat, and one child in the turn of the nineteenth century town of Sparta, Tennessee (1809).

1 The Pain of the Stone: Backgrounds of Urological Chemistry in the Nineteenth Century

Urological stones are painful. I would know, as I had one prior to a lecture in graduate school. I knew there was a problem when I experienced unexplained pain in my back, and in a really scary series of events, after toting books to the Thomas Cooper Library, passing blood in my urine. I immediately went to a series of doctors who agreed that different colored urines, especially a purple urine later on, were not good. After finding myself finally at the urologist, like uroscopies of old, commenting on what I could not see, with an X-ray instead of urine, Dr. John Fairey pointed out the small abnormality in the picture of my bladder, I had a urinary stone. He handed me a script for pain killers that he promised would keep me outside of the emergency room and it was going to pass on its own. Eventually, while preparing to help with a lecture at the University of South Carolina, I felt a pressure that I had to pass the stone. I ran after expelling the stone at the urinal to the nearest restaurant and got a soup cup to take the stone to the medical office for further analysis. Technicians there remarked at the number of spurs, or ridges, the stone had on its outside. Obviously, the more ridges a stone had the more painful it was to expel the stone. Through chemical analysis, the in-house medical pathologist revealed its chemical composition: calcium phosphate. Because of the knowledge of the chemical makeup of the stone, it was likely caused by my irritable bowel disease or the dehydration I had experienced living in the warm climate of Columbia, South Carolina. However, I was interested in the history of stones, as I had just happened to come across the writings of Edward Darrell Smith (1775/1776?–1819), a physician in Columbia during the nineteenth century.

Reading nineteenth century medical articles that I found accidently after becoming interested in the biography of Smith, who was a urologist, really intrigued me as a young graduate student. It was surprising that the pains of the stone, that Smith also suffered from, read just like my experience. Also, as a historian who had read some books about humoral theory, it was interesting that physicians like Smith, and also Thomas Cooper, whom the library was named after, were advocating a return to humoral theory ideas while also advocating for new chemical analysis,

E. A. Driggers, *Early Nineteenth Century Chemistry and the Analysis of Urinary Stones*, Perspectives on the History of Chemistry,
https://doi.org/10.1007/978-3-031-34973-7_1

much like the analysis that occurred with my urinary stone. Humoral theory was a very old idea of medical theory that might have originated with the Chinese, South Asian cultures, or the Ancient Greeks. Western medicine was very much inspired by the Ancient Greek theories of the humors. The Greeks thought that the body was made up of four key fluids that regulated and controlled the vital or life forces in the body. These humors included blood, phlegm, yellow bile, and black bile. All disease processes originated in their excess or deficient levels in the body. It was the physician's job to regulate those fluids in order to return their patient's body to health.

Urinary stones represented an interesting problem, as medico-chemists, or investigators at the turn of the nineteenth century who were interested in applying chemistry to medical problems, wanted to end the sufferings from urinary stones. The problem of the urinary stone also represented to me, a historian examining the problems of medicine and chemistry over 200 years later, a window to examine what happened to humoral theory as medicine changed and seemed to leave the theory behind, and move to more anatomical understandings of disease.

The United States epidemiological branch, the Centers for Disease Control and Prevention, located in Atlanta, Georgia, found in their National Health and Examination Survey that around eight percent of people worldwide will develop a urinary stone. In terms of defining a urinary stone, they are usually described, according to the Mayo Clinic, as being "... hard deposits made of minerals and salts that form inside your kidneys" [1]. There are many different types of stones and many different ways to classify the stones, such as by their physical make up or their chemical makeup. In many ways the development of collections for classification purposes of stones provide an end point for the book.

I was also born in the American South and found that I lived in what urologists call the "Stone Belt." The American South is the so-called stone belt because people living in the south were 50 more likely than other regions to experience a urinary stone [1–5]. I am more likely to develop a stone because culturally in the South, the Southern diet contains a lot of tannins in foodways like sweet tea, collard greens, black-eyed peas, squash, and other vegetables high in tannins. However, the temperature of areas also produces stones, such as hot areas in the South and even in other regions like California. Most urologists would propose that stone production is caused by high temperature, hydration issues, diet, disease, or in some cases pregnancy. In a call back to humoral medicine, dehydration causing stones is a literal fluid imbalance. Some physicians anecdotally advise patients to consume different fluids to combat stones, such as the importance of drinking lemonade or an occasional beer. Perhaps there are aspects of humoral medicine that continue to remain in medicine, especially in the case of treating and thinking about urinary stones.

1.1 History and the Problem of the Urinary Stone

Surgeons, physicians, and chemists saw bodily stones as an extremely difficulty problem. Stones, or calculi, could occur in many different parts of the body, such as in the mouth, bladder, lungs, brain, or could cause major problems in other parts of the body. For instance, America and Britain were white supremacist societies during the nineteenth century, and some physicians, like Benjamin Rush, believed that black skin color was a symptom of a disease, and thought internal blockages caused bodies to turn black, and that African slaves were simply sick white people.

The only known cure for calculi in any part of the body was its removal. The removal of the urinary stone was called a lithotomy. The term *lithos* was from the Greek for rock, and *otomy* mean removal or cutting. The early nineteenth century was also a time period without pain blockers or general anesthesia during surgeries, so a fast lithotomy was the goal of the surgeon and the patient. The pain of patients must have been excruciating to decide to go under the knife. Patients often complained about extreme pain in their legs, back, and in other areas of the body. Historically, we have patients writing about their experiences with lithotomies, even marking them as significant life events. The English Restoration writer and politician Samuel Pepys marked the day every year of surviving his lithotomy [6]: "Up early, this being by God's blessing the fourth solemn day of my cutting for the stone."

The surgeon and the physician wanted to prevent their patient from having to undergo such a painful procedure. Surgeons, though often proficient in performing the procedure, had to acquire much training and sometimes made errors in performing the operation. Physicians wanted to maintain their utility to patients suffering from the stone. However, both looked to chemistry to help them in their desires to better treat the stone, and perhaps to further understand disease. Shrinking stones could help surgeons extract the stone easier, perhaps without an operation, and allow them to maintain their professional prestige of being the provider that could remove the stone. Physicians hoped that a command of chemistry might allow them some type of intervention to prevent the patient from seeking the surgeon. Chemists, or those medical providers interested in chemistry, saw chemistry as a useful tool. William Hyde Wollaston, a successful chemist during the nineteenth century, had previously determined that chemicals, like solvents, could destroy bodily stones in the lab. But no one was sure of how to administer those solves into the human body. Chemist were excited that their practice could be extended to useful applications, like those of medicine and surgery. More knowledge about urinary stones could also allow chemists to explain more about the natural world.

While the application of chemistry was expanded to treating and understanding bodily stones, some physicians saw the new chemistry that was based in analyzing the body's fluids and substances, as a tool to advance a change in medicine: a return to humoral theory. The revised humoral theory, added to medicine and chemistry of the nineteenth century, was often called humoral pathology [7]. Some physicians were extreme in their desires to return to humoral theory. Wheeler framed the return as that [8], "humoral and climatic theory persisted in one form or another despite attempts to revise it or abandon it well into the nineteenth century." One point that is unclear in the historiography of humoral medicine is when it truly collapses and how it continues to periodically return. Historians like Delacy argued that the end of humoral theory resulted from it not being able to adequately understand the phenomena of contagion [7].

Understanding the return of humoral medicine in the attempts of nineteenth century surgeons, physicians, and chemists has a lot in common with explaining the changing notion of what a fever was at the turn of the nineteenth century [9]. This return to explaining urinary stones was necessitated by chemistry at the turn of the nineteenth century. But what was humoral theory at the turn of the nineteenth century.

1.2 Humoral Theory at the Turn of the Nineteenth Century

Humoral theory is a very old and ancient theory of medicine that has its roots in Middle Eastern and Chinese Medicine [10]. Humoral theory is often associated with ancient Greek medical systems. Disease was thought to be a misalignment or "imbalance" of one of the body's four humors. These humors consisted of phlegm, blood, black bile, and yellow bile. Eventually, systems of chemistry were added to each of the humors, such as complimentary four elemental system of earth, fire, wind, and water. By the nineteenth century, humoral ideas had changed and were not exactly like the ancient system. Humoral pathology was similar in the sense that it thought of disease as an imbalance of the humors, but what counted as a humor and what imbalance meant was up for debate. Humoral pathology, like humoral theory in different time periods, remained as separate ideas but lacked a coherent system. Since the time of Paracelsus, chemists and physicians had been refashioning the humors to suite different theories of disease. However, some surgeons, physicians, and chemists at the turn of the nineteenth century thought that diseases, including urinary stones and other bodily stones (they often referred to them as the morbid concretions), occurred through some irritation or imbalance of the fluids of the body. Therefore, studying the fluids of the body through chemical analysis might be a useful tool to understand diseases like bodily stones.

It is unclear how much resemblance the humoral pathology of the nineteenth century resembled the humoral theory of Galen and Hippocrates [11–14]. Hippocrates explained disease as not only humoral imbalances but also coming from changes in weather, airs, diet, or emotions temperament. Ideas in humoral theory continued to go through changes as they were received in Middle Eastern society, such as in the work of Ibn Sina (Avicenna), who added ideas like emotional temperament to the humors and ideas from Middle Eastern medicine and chemistry. Ideas like anatomical-based conceptions of disease and contagion challenged humoral theory, but never really fully dismissed the idea. Medical alchemists, which included Paracelsus, brought back essential ideas of the humoral back to medicine and chemistry, such as breaking all earthly substances into salts and earth, while Paracelsus hated any ancient medical authority [12].

Medical theories of balancing the body's fluids at the turn of the nineteenth century involved bleeding, blistering, purging, and cupping in the hopes to remove the irritation of the body and return it to an even status. Though the idea of balance and controlling the fluids of the body continued in nineteenth century medicine, the chemistry and medical theory underlying their importance waxed and waned [13–17]. Humoral theory did make sense to those who practiced it, as a fever was broken by breathing the vein (drawing off blood by a lancet), and the patient seemed to cool down and the redness from their face left [16]. Medical providers also practiced uroscopy, as well as fecalscopy sometimes, in order to determine what the nature of disease was in the body [18]. Often urine or other bodily substances that were determined to be irregular were taken to be signs of disease. Breathing the vein and uroscopy were certainly vestiges of humoral theory, but it was unclear exactly how tight the association physicians had with ancient medical practices. Often medical providers were adding new ways of chemical analysis to the fluids they were extracting from the patients. Discussion of bodily fluids with humoral terms were expedient for physicians, surgeons, and chemists, but it was unclear how tight those same providers clung to the ancient system, or associated what they were doing with a neo-humoral theory. However, some providers like Thomas Cooper, did not mix words about their project to have humoral theory return.

Discussing the nature of humoral pathology became a topic of conversation between physicians, chemists, and surgeons on both sides of the Atlantic. Providers and chemists living in the cities of Philadelphia, London, and Charleston debated and discussed humoral theories, along with ideas about how to help patients suffering from bodily stones. Using methods from Digital Huminites, the Google Ngram can reveal trends regarding publications, as the tool analyzes terms of publications that were scanned into google books. Figure 1.1 shows the trends in literature discussing the term "humoral pathology" in the years of 1820 through 1840. Humoral pathology had periods of popularity, and it is in that time of popularity that this book is set.

The Ngram results might indicate a time of popular debate among physicians, chemists, and surgeons in regards to humoral pathology. The Ngram also reveals periods of popularity about the writing about urinary calculi and humoral pathology, as noted in Fig. 1.2.

The new chemistry that was developed in the Enlightenment inspired a renewed interest in humoral pathology at the turn of the nineteenth century in the United Kingdom and in the United States. Though there was a lot of interest in chemistry in studying and treating bodily stones that included urinary calculi, there was only one effective treatment for the stone: surgery. However, the lithotomy was very painful.

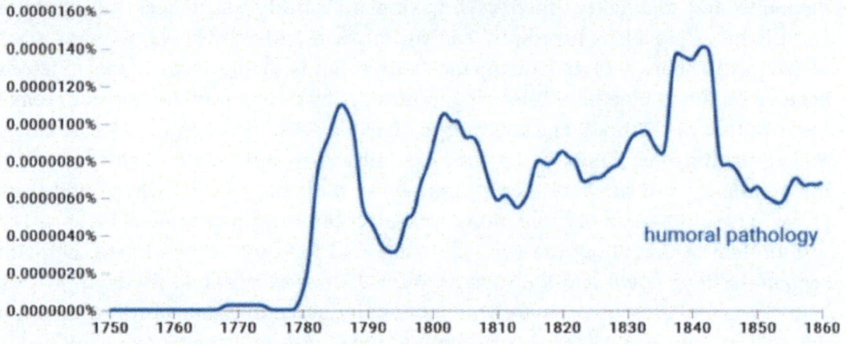

Fig. 1.1 Google Ngram analysis of "humoral pathology" This graph shows the confluence of when terms like humoral pathology hit a high popularity, such as during the time period of this book

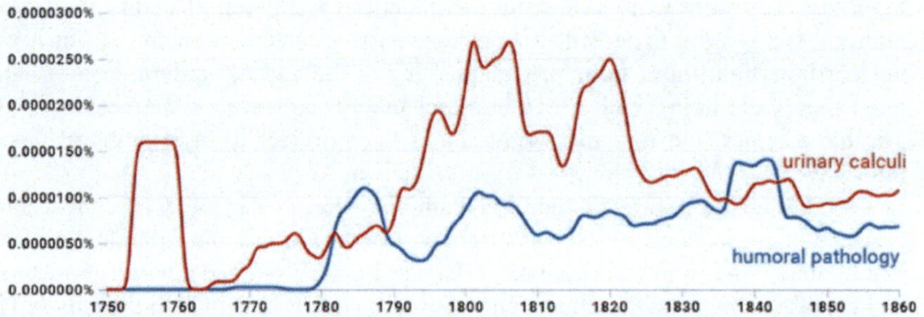

Fig. 1.2 Google Ngram analysis of "humoral pathology" and "urinary calculi". The terms urinary calculi and pathology are compared in regards to their uses among the digitized google books so far. Note that the terms hit a high point during the time period of this book

1.3 Pain and Lithotomies

As the reader can tell, the operation for the lithotomy (Fig. 1.3) is beyond painful and we can only imagine how terrible the operation was for the patient, as there were no pain killers or anesthesia. Surgeons saw the operation as a point of professional pride, but also knew the risks, as patients often died. Surgeons also had information about how many patients died during the lithotomy. In the article "An Analysis of Statistics of Lateral Lithotomy," which pointed out the information the surgeon William Keith was able to gather, about 20% of patients died in the operation. Keith noted that out of 221 cases about 45 patients died [19]. Alexander Marcet, a physician and chemist who appears later in this book, studied in great detail how deadly the lithotomy was for patients and publishes his results. The information about death motivated surgeons, physicians, and chemists to look for tools to either improve or eliminate the need for the lithotomy.

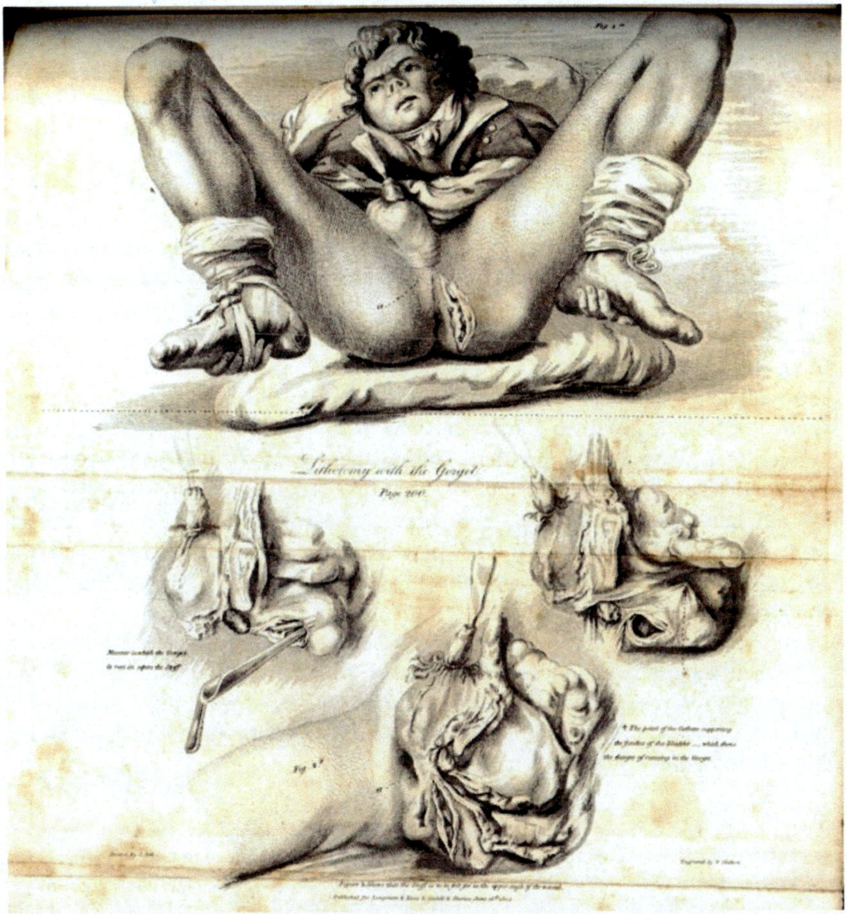

Fig. 1.3 "Posture for lithotomy" and "lithotomy with the gorget," from John Bell's the principles of surgery (1806), Reynolds-Finley historical library, the university of Alabama at Birmingham, with permission

Surgeons had a strong desire to improve lithotomies. One effort at improvement was taking detailed case information of patients suffering from the stone, and even collecting the stones to train others. Surgeons prided themselves in continuing to learn and improve their own practice. Keith wrote in that same article about deaths in lithotomy that, "As we and others often derive most instruction from our failures, I shall, therefore, briefly give the leading points connected with the fatal cases of my lithotomy experience [19]." Perfecting lithotomies, for surgeons was about continual education and seeing out feedback from other surgeons [20]. Surgeons also thought that publishing information or case histories of the procedure would improve the community as a whole and train student with little experience. This was a definite change from the secretive and proprietary procedures of surgery in the time of Frère Jacques [21].

Edward Stanley, who was an assistant surgeon and also lecturer in anatomy and physiology at the St. Bartholomew's Hospital in London published in 1829 *An Account of the Mode of Performing the Lateral Operation of Lithotomy* [22]. Stanley wanted to both improve the medical school and the museum at St. Bartholomew. His work anatomical work won him both the Hunterian Prize and the Jackson Prize. He was born in 1793 and was trained at the St. Bartholomew's Hospital, where he is credited with opening up the museum there; both the museum and the catalogue received positive attention from peers [23]. Stanley had a good reputation as a teacher and he severed in 1843 as the President of the Royal Medical and Chirurgical Society, and then in 1858, he served Queen Victoria. He passed away while viewing a surgical operation in 1862 of what Sir William MacCormac called a "cerebral hemorrhage" [23]. However, not all his colleagues spoke kindly of him, as William Lawrence, who is credit with pro-evolutionary ideas, was critical of Stanley as he died. During the operation that Stanley was viewing on the day of his death, felt the pain of a seizure, which he remarked was not anything of note. However, as Stanley was dying from that seizure, Lawrence remarked to Stanley that [23]: "Ah, poor Stanley, wrong to the last." Other colleagues described Stanley in more glowing terms [23]: "... 'He was a blunt, kindly, humorous, straightforward, and honest man.'"

An Account of the Mode of Performing the Lateral Operation of Lithotomy was published version of the lecture that Stanley gave as the Hunterian lecture. Much of the emotion regarding the lithotomy comes out in the work [22]. Surgeons also had anxiety regarding the operation, as he included a story about another surgeon, Mr. Cheselden, recording the anxiety of Cheselden [24]:

> If I have any reputation in this way, I have earned it dearly; for no one ever endured more anxiety and sickness before an operation; yet, from the time I began to operate, all uneasiness ceased; and if I have had better success than some others, I do not impute it to more knowledge, but to the happiness of a mind that was never ruffled or disconcerted, and a hand that never trembled during an operation.

Surgeons were under immense stress and likely experienced anxiety around the procedure, with Stanley making the link that if experienced surgeons such as Cheselden have these feelings, it would be extremely difficult for an inexperienced surgeon or student would have a tremendously difficult time performing the operation for the first time. Stanley wanted to write about the operation in order to ease students and first time practitioners into the operation as he had not found another detailed account of the operation [24]: "... I have experienced the want of a simple, yet sufficiently detailed account of the mode of performing the operation, unencumbered by critical or historical matter." He also includes pictures to help surgeons learning the lithotomy.

Stanley meticulously sketches out for the reader the nature of the operation. The surgeon will start in the perineum, which is located between the patient's urological organs and the anal area. The surgeon will start in this area with their first cut, which the surgeon will continue to cut until they reach the patient's prostate. Stanley recommends this approach because it provides extra protection to the patient, as it avoids the intestines or cutting into fecal matter. Stanley disagrees with Cheselden in whether the patients should completely void their bladder, Stanley was of the opinion that the patient should have a small bit of urine so that it will indicate the surgeon that they have reached the bladder. The forceps are recommended in the removal of stones from the body.

Stanley then shifts to the next steps of the operation. The surgeon should then utilize the staff, which was a slender piece of metal that was inserted in the bladder by the surgeon, or assistants, during the lithotomy. The surgical assistant had to hold the area that had been opened [25]: "The straight part of the staff should be held perpendicularly: its curved part should not be pressed in one or other direction against the sides of the urethra." The assistant then used the staff to keep the cut area open, and with their hands they remained on the right side of the operating table [26]: "That the staff should be held steadily, in the position in which it is received from the operator, is indispensable." While the surgeon was operating, the assistants had to hold down the patient and keep them from moving [27]:

> Beside the assistant holding the staff, two other assistants, at the sides of the patient, by firmly grasping his knees and feet, must resist the efforts he will make to raise and approximate his knees, much to the embarrassment of the operator. Another assistant should be placed at the patient's shoulders. The pelvis, being brought to the edge of the table, is to be so adjusted that the anterior superior spines of the ilia will be in a horizontal line, and it is to be kept in this position until the operation is completed.

The next step of the operation was to make a cut, "an inch and a quarter above the anus" [27]. The surgeon then takes his index finger, as well as his thumb, which reminds the surgeon where he should end his cut, and places it in position on the patient. The cut should be about three inches in total. The scalpel is then used by the surgeon to make a cut into the urethra. Next the gorget is used, which is a surgical tool that was also inserted into the bladder, and this often causes urine to come out the cut. The gorget has a head that resembles a funnel. Then the surgeon has a decision to make: does he continue to use the knife or the gorget. Stanley recommends using the finer and inserting it to the bladder, however, sometimes the surgeon will have to move the prostate out of the way to continue the operation. Often times forceps are used to search for the stone in the bladder, or some combination of the knife or staff.

Stones are often removed with the forceps. Or sometimes fingers are used to manually remove the stone. Water or other substances might be used to remove the stone. Stanley also emphasized the importance of assistants in the operation, as they were often surgeons in training, that helped the surgeon, but also kept the patient from resisting during the operation. The assistants also helped the patient recover from the surgery [25]: "An operator, without the advantage of experienced assistants, must himself ascertain that the bandages are applied in a manner to prevent the possibility of their yielding." This description from Stanley's work should show the pain of the lithotomy and explicate how important it was for chemists, physicians, and surgeons to either make the stone easier to remove through chemicals or prevent the stone from forming.

1.4 Chemical Medicine

The description of the lithotomy would inspire fear and pain in anyone. It is clear why surgeons, physicians, and chemists wanted to study bodily stones in order to prevent people from suffering from them. They wanted a better way to treat the stone than the lithotomy, or at the very least, make the operation easier. First, there needed to be a mastery of how stones formed in the body. Physicians wanted to analyze stones in order to both advance their own chemical medical programs (these physicians could be considered medico-chemists) but also wanted to advance knowledge in their communities, including medical communities that are interested in humoral pathology. John Harley Warner in *The Therapeutic Perspective* found that physicians were excited about scientific ideas prior to understanding how those scientific ideas would be utilized in clinical practice. How physicians were interpreting scientific ideas was continually in flux, causing Warner to conclude that, "Once we are free from looking for the *one* meaning of science in medicine, then the work of the past two decades appears not merely as a fragmentation of perspectives in need of integration, but as a springboard to posing more refined questions" [28]. Science in American Medicine was not a singular entity [29]:

> ... [S]cience' in American medicine was never monolithic. It meant different things at various times and to assorted social groups, including the varieties of medical practitioners ... a self-conscious awareness of the multiplicity of the meanings of science in medicine is one of the most promising guides to research that can illuminate the place and function of science in medicine and of medical science in American culture.

Humoral pathology was science at the turn of the nineteenth century. This idea was especially true in certain communities of medico-chemists, such as Edward Darrell Smith and Thomas Cooper, who are discussed in this book.

The changes in humoral medicine could be analogized to the work of Lawrence Principe in his book *The Secrets of Alchemy* [30]. In the book he discussed that though alchemy and chemistry changed during the Chemical Revolution, alchemy continued to be practiced, though differently from ancient and medieval alchemy. People practiced alchemy into the early twentieth century, and there are several examples of alchemical practice today. One example can be found with Berkley chemists turning lead into gold, or transmutation, though it was a very small amount [31]. Humoral pathology was involved in larger changes in chemistry and medicine, and though the chemistry that underlined humoral pathology evaporated in the wake of the chemical revolution, the new revolutionary chemistry was used by some medico-chemists to prop up the idea. They switched one chemistry for another. But what were these revolutionary changes in chemistry.

1.5 Revolutions in Chemistry

After teaching the chemical revolution in my classes for several years, I feel that the concept is still elusive and there is a lot of room for scholarly debate, and clearer explanations are needed for students. Many readers, especially practicing physicians, are here to discuss urinary stones; however, I anticipate that many of the chemistry people are interested to hear my summary of the chemical revolution. Also, both audiences need to understand how important bodily stones are to the chemical revolution.

The chemical revolution, which is thought by many scholars to occur around the late eighteenth century, and it was important in the revisions occurring in humoral pathology. The chemical revolution is thought by many scholars to consist of several transitions which prioritized mathematical and precisions, often with instrumentation, and "chymists" moved away from the secretive and clandestine alchemical practices. Though everyone knows the example of Isaac Newton continuing to practice alchemy secretly [32].

The focal point for a lot of scholars in regards to the chemical revolution is the French tax collector turned chemical investigator Antoine Lavoisier (1743–1794). Lavoisier brought instrumental precision to analyzing chemicals and substances. The close analysis of Lavoisier brought about the idea of reducing chemicals into their smallest components, Lavoisier thought of these small parts of substances as elements. Lavoisier achieved a lot with being able to break down substances, as he was able to break down water, a substance that many alchemists and chemists of the time thought it was an irreducible substance or even a primary substance [33]. He was able, through the use of electricity, to break down water into the "elements" of hydrogen and oxygen [34].

Lavoisier was building on the work of other chemist, such as Robert Boyle, known to many readers as the discover of Boyle's law regarding gasses. Also, there are a lot of contributions of women to the chemical revolution that have not been properly addressed [35]. However, Lavoisier's work in the chemical revolution helps with explanations because he gives the movement both a textbook and terminology.

Prior to Lavoisier's revolutionary chemical work, chemists were examining stones of the body. Joseph Black (1728–1799) a medico-chemist published a doctoral thesis while at the University of Edinburgh on urological calculi. Black was looking for chemical substances that could dissolve urinary stones. Black was part of a movement in Scotland regarding the application of chemistry to practical problems that John Stewart and other scholars have examined [36]. Though Black was looking for a substance that dissolved stones, he could not find one that would be safe for human consumption. While undertaking his chemical research for stone solves he found out more information about gasses and different types of "airs [37]." Medico-chemists had known that certain acids would dissolve stones but they had yet to find substances that were safe to digest. Medico-chemists like Black wanted to find out more about substances that dissolved stones in the hopes that their application could help patients, but also teach chemists more about the natural world.

Lavoisier published his *Elements of Chemistry* in 1789, with the English edition appearing in 1802 [38]. In the original *Elements*, Lavoisier examined urine with his rigorous chemical analysis. The 1802 Kerr edition also added bladder stones and biliary stones (from the intestines) to the book as well. After studying Lavoisier's techniques, Kerr was interested of analyzing the fluids of the body more deeply, such as blood and bile, as well as other substances such as milk and cheese [38]. Kerr emphasized the importance of decomposition, or analyzing substances down to their smallest parts (elements) [39]: "By chemical analysis, they [biliary concretions] afford resin of bile, benzoic acid, and small quantities of lime, soda, and neutral salts, having a basis of ammonia." Kerr and other medico-chemists embraced the chemical methods of Lavoisier because they were excited that it was the ultimate answer to the diseases of the body. Kerr was writing in the time of the Enlightenment, where scholars looked for a universal key to unlock all knowledge [40]. Unfortunately, Lavoisier was ultimately executed during the Terror of the French Enlightenment in 1794, his ideas about chemistry continued to live and were exported to England [34].

Other scholars interested in Lavoisier's ideas continued to apply them to the fluids of the body and continued the program of analyzing the body's fluids. In France, Antoine Francois, Comete de Fourcroy, and Louis Vauquelin examined human urine to understand its relationship to urinary stones [41]. In England, physicians like George Pearson popularized and translated Lavoisier and extended his chemical ideas to urine. Pearson published work on the analysis of urine in publications like *Philosophical Transactions*. He claimed to find a link between arthritis and urinary problems: they were caused by conceptions in the body [42–44].

Other chemists were interested in applying Lavoisier's methods to understanding the fluids of the body, such as the Copley Medal Winner, William Hyde Wollaston, who served in leadership in the English scientific organization: The Royal Society. Wollaston continued the investigations into the chemical concretions of the body and their related diseases, such as gout. A transition occurred in the chemical revolution, as chemists and their ideas fled the 1789 French political Revolution, especially when the Revolution turned violent, and found homes in England, and in some cases America.

1.6 The Chemical Revolution Spreads to England and America

The French Revolution, especially its violent aspects, spread people and ideas from France to England and America. Many of the characters in this book, including Alexander Marcet, were affected by the Revolution, as Marcet had to flea Genova to avoid becoming a soldier in the resulting international wars of the Revolution. The violent nature of France, especially in the time of Robespierre, caused surgeons, physicians, chemists, and other intellectuals to flea France to London and other American coastal cities like Charleston, South Carolina and Philadelphia Pennsylvania. Many French intellectuals were attracted to America because of their perception of its freedom and democracy. Catherine A Herbert wrote of the attraction of chemists and other intellectuals leaving France for Philadelphia [45]: "A motley group of émigrés and other refugees from France, settled in Philadelphia, attracted by the capital's cosmopolitan reputation, its conception of important American personalities, and the promise of helpful connections and potential financial rewards." Other French immigrants would bring the principles of Enlightenment chemistry to America, and contribute to chemical knowledge in Philadelphia. French immigrants also traveled from the Caribbean to Charleston, in order to flee the Napoleonic Wars and later the Hattian Revolution [46].

Surgical knowledge about urological stones continued to be published in French, but chemical work analyzing urinary stones expanded into English language publications. English museums also began collecting stones from surgeons, which provided a large source base for people like Alexander Marcet to analyze and run chemical studies. Chemists in the United States, as well as the United Kingdom, wanted to take the Lavoisier chemistry and apply it to several problems of stones blocking the body's fluid systems. Though this study focuses on English language sources, more studies about urinary stones and chemistry needed to be performed on the French medical sources at the turn of the nineteenth century as well.

1.7 Medicine Prior to the Period of Professionals

This study takes places prior to the popularization of the theory of germs [47]. However, the fluids of the body are part of the story of understanding disease, as those studies also shifted to the microscopic. For instance, Golding Bird, a British chemist, started examining urinary stones by writing of their crystal-like nature [48]. Bird's urological work is an example of the medical transition from seeing diseases processes holistically, to more of a focus on the isolated parts of disease. This book takes places in the time of some physicians trying to take aspects of ancient medical ideas and putting them together with new chemical ideas [49]. Sometimes older medical ideas return with the addition of new scientific approaches to knowledge.

Another way of thinking about the history of medicine comes from the history of science, specifically in the work of Andrew J. Lewis. Lewis studied the transition in science from amateurism to professionalism [50]. Lewis saw a change in American science, where anyone could participate in the making of scientific knowledge in the early eighteenth century, as many people of many different roles in society contributed to scientific journals, collected fossils, and contributed to scientific ideas. There was a shift in science as questions required more sophisticated and expensive scientific instruments, such as the work of physics in the late nineteenth century, but chemistry could also be added to this description. It was a shift, Lewis claimed, from a "democracy of facts" to "an Empire of Reason" [50]. He framed the change [51]:

> Ordinary Americans and naturalists themselves were asking similar questions: Just what was natural history for? What exactly did it do? Who should be considered a naturalist and why? Early republic Americans agreed that natural history could catalogue and describe nature, but, as practiced by an orthodox elite in the United States, it imposed limitations on theorizing about nature's cases and emphasized the assembly of individual facts.

With this transition in science, amateurs are replaced by professionals who exclusively can produce knowledge. A similar change happens in medicine, as medical knowledge creation becomes exclusive to only one type of professional practitioners.

In the cities of London, Philadelphia, and Charleston, medical care was performed by a variety of providers and the patient had a marketplace of options [52]. The members that participated in medical societies and related groups in science were made of many providers, intellectuals, and philosophers [53]. And societies made up of all types of providers published journals and other printed sources to communicate their ideas, which many different types of persons read. The printed materials were communicated to other cities in American and across the Atlantic, in a type of medical and scientific "Republic of Letters" [54]. But the contributions to medical and scientific journals are almost impossible to the average American or British person.

Along with the pushing out of the amateur of journal literature, but also the pushing out of providers of medical care, especially in America. Scholars are divided on the nature of the project of the professionalization of medicine. Scholars like Starr see the project very cynically, as physicians from universities wanted to control medicine in order to increase profits, while scholars like John Harley Warner argued that physicians wanted to elevate medical practice through the increase of scientific methods, especially those from Paris [28, 55]. However, this book takes place before major professionalization moves in American and in Britain [56].

Examining the questions related to morbid concretions, urinary stones, and other bodily calculi reveal a lot about the collaboration between medicine, surgery, and chemistry prior to the siloing in the professionalization of science and medicine. The nature of what a doctor, surgeon, or chemist did at the turn of the nineteenth century was not as disciplinary bounded as it is today, and their scope of practice were potentially large, as you will see in this book. Also, their training and entrance into their practice varied widely. In some cities there were organization that trained and contributed to community building, such as the Royal College of Surgeons. Some physicians had universities degrees and some did not. Chemists taught themselves the principles of Lavoisier and other ideas in early teaching systems of chemistry or some learned through industry or some type of apprenticeships. Surgeons and physicians often received additional training through formal and informal apprenticeships [28, 57].

As mentioned above, the boundaries between chemistry, surgery, and medicine were not as clear as those today, nor were they sharp at the turn of the nineteenth century. This book examines the work of chemists, surgeons, and physicians in the same vein of Susan Leigh Star and James R. Griesemer, who noted that special objects seems to bring disciplines together and unite them in common questions [58]. Stones of the body (morbid concretions) like urinary stones brought together surgeons, chemists, and physicians around the desire to understand why stones formed and whether anyone could do anything to treat them, or prevent them from forming.

1.8 A Plan of Action

This book attempts to understand bodily stones in their chemical, medical, and surgical contexts in America and the United Kingdom, specifically in the cities of London, Philadelphia, and Charleston. In this book, humoral pathology in its new chemical context is explored around these same turn of the nineteenth century cities regarding its application to understanding bodily stones. This book also shows that humoral theory stayed around in the medical consciousness of some medical providers, surgeons, and chemists. Many of the historical concepts of these ideas have been introduced in this first introductory chapter.

Chapter 2explores morbid concretions and their role in medical and scientific practice in both the United States and Britain. This chapter will focus on the nature of humoral pathology during the early nineteenth century and how it was utilized in medicine, even though it was not a full acknowledged part of comprehensive practice. Medicine will be contextualized with a short exploration of changes in nineteenth century chemistry and their application to morbid concretions and the general chemistry of the body.

Chapter 3 starts the main body of the book, as it explores individuals and their theories about urinary stones and stones of the body. Chapter 3 examines the life of Benjamin Rush who was one of the first doctors to teach chemistry at an American university, which was Penn. Rush received extensive chemical training under Joseph Black while in Scotland, and received additional chemical training by interacting with other British chemists. He was a professor of chemistry and eventually medicine at the University Pennsylvania in Philadelphia, Pennsylvania. Rush was passionate about some concepts of humoral pathology and his chemical training paired well with that perception of medicine. He thought of diseases as being caused by problems with the fluids of the body. Though Rush did not support slavery, he subscribed to the idea that African slaves were ill because of their abnormal black colored skin. He, like many nineteenth century chemists, looked for chemical medicines to cure the disease they suffered from that manifested in their skin. Rush also used chemical remedies to treat worms in the body and humoral interventions like blood letting to treat urinary stones. Rush's work shows of humoral theory, chemistry, bodily stones, and problems with the body's fluids were all linked with disease processes.

Chapter 4 removes the book's focuses from major metropolitan areas like Philadelphia and changes focus to rural America and the emergence of new cities in areas that were formerly backcountry, like Columbia, South Carolina. Edward Darrel Smith, a Charleston born, but Philadelphia educated physician, who later becomes interested in chemistry, studied and published ideas about how to treat urinary stones and speculated on their humoral causes. Smith's life illustrates how chemical ideas were moving from the broader Atlantic to the Carolina backcountry; but those ideas were also being sent back to the Atlantic World through periodicals [59].

Chapter 5 explores the rhetoric of chemistry and its relation to medicine, especially in the case of Thomas Cooper. Cooper, one of the most vocal and extreme advocates of humoral pathology, was also a professor of chemistry at the newly formed South Carolina College, after himself being a political immigrant from Europe. Cooper saw the fluids of the body as the key to causing disease and diseases occurred when those bodily fluids were blocked. Cooper makes his strongest arguments for his humoral framework and the importance of chemistry in understanding disease in his *Discourse on the Connexion Between Medicine and Chemistry*. However, Cooper was a controversial figure in American history as he also had differing views on religion and advocated for South Carolina to succeed from the Union.

Chapter 6 examines the relationship between surgery and chemistry, with the examination of the friendship between the surgeon Sir Astley Cooper and chemist/physician Sir Alexander Marcet. Cooper contributed many surgical calculi specimen to Marcet, who published a chemical analysis of several surgically removed stones in his 1817 *An Essay on the Chemical History and Medical Treatment of Calculous Disorders*. The two men had aligned interests as a surgeon wanted to know about potential treatments to help in lithotomies, while Marcet wanted to analyze, categories, and perhaps derive general principles about the chemical nature of calculi in the body. Marcet also gathers statistical information about the occurrence of stones, but also the success of lithotomies.

Chapter 7 takes a comparative approach, as the chapter examines the medicochemical workings of the examination of bodily stones in Philadelphia, Charleston, and London. The chapter examines several medical and chemical societies that were in operation in the cities and historicize the member's work on bodily stones. The chapter also includes museums like the Royal College of Surgeon's Hunterian museum and its increasing collections of urinary stones. It was in these societies and museums that surgeons, physicians, and chemists collaborated in their community, and even with other communities, in order to understand the problem of the stone. And Chap. 8 provides a conclusion to this book.

References

1. Mayo Clinic (2021) Kidney stones. https://www.mayoclinic.org/diseases-conditions/kidney-stones/symptoms-causes/syc-20355755. Accessed 3 Feb 2021
2. Dallas KB, Conti S, Liao JC, Sofer M, Pao AC, Leppert JT, Elliot CS (2017) Redefining the stone belt: precipitation is associated with increased risk of urinary stone disease. J Endourol 31:1203–1210
3. Clayton DB, Pope JC (2011) The increasing pediatric stone disease problem. Ther Adv Urol 3:3–12
4. Perry R (2016) More southerners at risk for kidney stones. Tennessean May 8
5. Eisner BH, Sonalia S, Herrick B, Pais VM Jr (2012) The effects of ambient temperature, humidity and season of year on urine composition in patients with nephrolithiasis. BJU Int 110:E1014–E1017
6. Pepys S (1899) Diary of Samuel Pepys. In: Bright M et al (eds) George Bell & Sons, Cambridge, p 210
7. Delacy M (2016) The germ of an idea: contagionism, religion, and society in Britain, 1660–1730. MacMillan, New York
8. Wheeler R (2000) The complexion of race: categories of difference in eighteenth-century British culture. University of Pennsylvania Press, Philadelphia, p 24
9. Hamlin C (2014) More than hot: a short history of fever. Johns Hopkins, Baltimore
10. Craik EM (2009) Hippocratic bodily "channels" and oriental parallels. Med Hist 53:105–116
11. Hippocrates (1978) Hippocratic writings (trans: Chadwick J). Penguin Books, New York
12. Sirasi NG (2013) Communities of learned experience epistolary medicine in the renaissance. Johns Hopkins Press, Baltimore
13. Sirasi NG (2007) History, medicine, and the traditions of renaissance learning. University of Michigan Press, Ann Arbor
14. Porter G (2007) The greatest benefit to humanity. W. W. Norton, New York
15. Jones A, Taub L (eds) The Cambridge history of science: ancient science. Cambridge University, Cambridge
16. Siraisi NG (1990) Medieval and early renaissance medicine: an introduction to knowledge and practice. University of Chicago Press, Chicago
17. Pagel W (1986) From Paracelsus to Van Helmont. In: Winder M (ed). London, Variorum Reprints
18. Stolberg M (2015) Uroscopy in early modern Europe. Routledge, Oxford, p 2015
19. Keith W (1871) An analysis of statistics of lateral lithotomy. British Med J 2:320–322
20. Gelfand T (1980) Professionalizing modern medicine: Paris surgeons and medical science and institutions in the eighteenth century. Greenwood Press, Westport
21. Ganem JP, Carson CC (1999) Frère Jacques Beaulieu: from rogue lithotomist to nursery rhyme character. J Urol 161:1067–1069
22. Stanley E (1829) An account of the mode of performing the lateral operation of lithotomy. Longman, Rees, Orme, Brown, and Green, London
23. MacCormac W (1900) An address of welcome by Sir William MacCormac President of the Royal College of Surgeons of England. Ballantyne, Hanson, & Co., London, pp 109–111
24. Stanley E (1829) An account of the mode of performing the lateral operation of lithotomy. Longman, Rees, Orme, Brown, and Green, London, pp preface
25. Stanley E (1829) An account of the mode of performing the lateral operation of lithotomy. Longman, Rees, Orme, Brown, and Green, London, p 3

26. Stanley E (1829) An account of the mode of performing the lateral operation of lithotomy. Longman, Rees, Orme, Brown, and Green, London, p 4
27. Stanley E (1829) An account of the mode of performing the lateral operation of lithotomy. Longman, Rees, Orme, Brown, and Green, London, pp 4–5
28. Warner J (1986) The therapeutic perspective: medical practice, knowledge, and identity in America, 1820–1885. Harvard University Press, Cambridge
29. Warner J (1995) The history of science and the sciences of medicine. Osiris 10:164–193
30. Principe L (2013) The secrets of alchemy. University of Chicago, Chicago
31. Lynch C (2021) Turning lead into gold: The Alchemists' goal of nuclear transmutation has been achieved, but not in the way they thought. Discover Magazine: https://www.discovermagazine.com/the-sciences/turning-lead-into-gold
32. Newman WR (2018) Newton the alchemist: science, enigma, and the quest for nature's "secret fire." Princeton University Press, Princeton
33. McEvoy JG (2010) The historiography of the chemical revolution. Routledge, New York
34. Donovan A (1993) Antonine Lavoisier: science, administration and revolution. Cambridge University Press, Cambridge
35. DiMeo M (2021) Lady Ranalagh: the incomparable life of Robert Boyle's Sister. University of Chicago Press, Chicago
36. Stewart J (2013) Chemical affinity in eighteenth-century Scottish physiology and agriculture. PhD Dissertation, University of Oklahoma
37. Levere TH (2001) Transforming matter: a history of chemistry from alchemy to the buckyball. Johns Hopkins Press, Baltimore
38. Lavoisier AL (1802) Elements of chemistry (trans: Kerr RW). Creech, Edinburgh
39. Lavoisier AL (1802) Elements of chemistry (trans: Kerr RW). Creech, Edinburgh, p 214
40. Maxwell N (2019) Science and enlightenment: two great problems of learning. Springer, New York
41. Leicester HM, Klickstein HS (eds) (1952) A source book in chemistry, 1400–1900. Harvard University Press, Cambridge
42. Pearson G (1798) Experiments and observations, tending to show the composition and properties of urinary concretions. Philos Transitions R Soc 88:15–46
43. Coley NG (2003) George Pearson MD, FRS (1751–1828): "the greatest chemist in England." Notes Records R Soc 57:161–175
44. Pearson G (1810) Observations and experiments on pus. Philos Trans R Soc 100:294–317
45. Herbert CA (1984) The French element in Pennsylvania in the 1790s: The Francophone immigrants' impact. Pennsylvania Mag Hist 108:451–469
46. Blaufarb R (2005) Bonapartists in the borderlands: French exiles and refugees on the Gulf Coast, 1815–1835. University of Alabama Press, Tuscaloosa
47. Worboys M (2000) Spreading germs: diseases, theories, and medical practice in Britain, 1865–1900. Cambridge University Press, Cambridge
48. Bird G (1859) Urinary deposits: their diagnosis, pathology, and therapuetical indications. Philadelphia, Blanchard & Lea
49. Rosenberg C (1979) The therapeutic revolution: medicine, meaning, and social change in nineteenth century America. In: Vogel MJ, Rosenberg C (eds) Therapeutic revolution: essays in the social history of American medicine. University of Pennsylvania Press, Philadelphia
50. Lewis AJ (2011) A democracy of facts: natural history in the early republic. University of Pennsylvania Press, Philadelphia
51. Lewis AJ (2011) A democracy of facts: natural history in the early republic. University of Pennsylvania Press, Philadelphia, p 1
52. Starr P (1982) The social transformation of American medicine. Basic Books, New York
53. Fothstein WG (1985) American physicians in the nineteenth century: from sects to science. The Johns Hopkins University Press, Baltimore
54. Goodman D (1994) The republic of letters: a cultural history of the French enlightenment. Cornell University Press, New York

55. Warner JH (1994) Against the spirit of the system: the French impulse in nineteenth-century American medicine (2003). Johns Hopkins University Press, Baltimore
56. Bonner TN (1995) Becoming a physician: medical education in Britain, France, Germany, and the United States, 1750–1945. Oxford University Press, New York
57. Freidson E (1988) Profession of medicine: a study of the sociology of applied knowledge. University of Chicago Press, Chicago
58. Star SL, Griesemer JR (1989) Institutional ecology, "translations" and boundary objects: amateurs and professionals in Berkeley's museum of vertebrate zoology, 1907–39. Soc Stud Sci 19:387–420
59. Delbourgo J, Dew N (eds) (2008) Science and empire in the Atlantic world. Routledge, New York

2 No Stone Unturned: The Chemistry of Morbid Concretions

In 1835, John Bostock, a chemist and physician, read "On the Chemical Constitution of Calcareous Tumours of the Uterus" to the Medico-Chirurgical Society of London [1]. The Medico-Chirurgical Society was made up of pharmacists (apothecaries), physicians, surgeons, and chemists from across metropolitan London who shared interests in questions related to science and medicine. Bostock worked with chemists and physicians on questions related to the fluids circulating in the human body. Throughout his career he worked with chemists like Thomas Thomson and Jöns Jacob Berzelius. He read to the audience that [1]:

> It is well known that various parts of the body are liable to have deposits formed in them of an earthly or bony matter. These bodies have, in a few instances, been made the subject of experiment, and have been found generally to be composed of the phosphate of lime, combined with a small portion of the carbonate cemented together by a quantity of animal matter.

Bostock's paper was an exploration of how concretions in the body could be understood chemically, and he hoped that by publishing papers on this topic, this would improve knowledge about how to end concretions in the body. Bostock, like other physicians, surgeons, and chemists across Britain, wanted to study bodily stones with the new chemistry in order to reduce these concretions down to their base substances, much like Lavoisier did in his water experiment in defining what an element was in the new chemistry. Hope extended to many other diseases beyond concretions like urinary stones, but conditions like gout, diabetes, and even the case of uterine concretions that Bostock presented to the Medical-Chirurgical Society.

Medico-chemists, or physicians interested in chemistry often wrote about their means of collecting concretions as well. In Bostock's paper he worked with a Dr. Richard Lee, a fellow member of the Medico-Chirurgical Society. Lee also had training as midwifery as well as medicine. Like the Alexander Marcet and Astley Cooper relationship mentioned in a future chapter, one man was able to gather

E. A. Driggers, *Early Nineteenth Century Chemistry and the Analysis of Urinary Stones*, Perspectives on the History of Chemistry,
https://doi.org/10.1007/978-3-031-34973-7_2

stones for the other to put to chemical analysis. Lee presented Bostock with seven uterine calculi from patients that had been under his care, to which Bostock put under chemical analysis and described to the reader: all seven were of different size and all had some white, yellow, or ivory coloring. Bostock applied heat, after which he proceeded to measure them, and then exposed them to different acidic reagents.

Physicians were building on existing chemical work that was going on in animal chemistry, a field of chemistry that studied the living processes of the body. The chemistry of animals and plants was the beginnings of what chemists consider organic chemistry. Chemists in Britain were studying many diseases through their research, from urinary stones, gout, diabetes, and diseases they thought were related to the fluid blockages of the body. Chemists and physicians had a shared interest in understanding the chemistry of the body. Chemists wanted to extend their research into the deepest principles of the body's function to show how important their new chemistry was to the world, while physicians wanted to offer their patients better healing powers. Surgeons were also intrigued, as better chemistry could improve the lithotomy.

The early nineteenth century was a time period of British chemistry coming of age, as the emphasis on chemical knowledge production changed from France to Britain because of the political instability of France and the death of Lavoisier in the terror [2]. Some chemists were also fanning the flames of nationalism in chemistry, such as Alexander Marcet's praise of the work of William Hyde Wollaston [3]:

> The large, dismal, subterraneous laboratory of olde chemists, is now changed for the fire-side of a comfortable study; and a new school is rising under his auspices, and those of two or three other British chemists, which promise to give a most essential impulse to the progress of analytical chemistry.

Nationalistic sentiments were expressed in the work of Thomas Thomson, another British chemist, in his *System of Chemistry*, which was designed to teach chemistry [4]. Thomson wrote of the popularity of British chemistry, and he was optimistic about the future of chemistry in Britain. Though there was a growing national pride of chemists in their potential and their rising popularity, some of the practitioners, like Alexander Marcet, were from other nations who adopted a British identity.

Animal chemistry was seen by those seeking to study the stones of the body as a proper tool to further the understanding of disease. This chapter is a study of how animal chemistry was brought to study the fluids and concretions of the body by physicians and chemists. The inspiration of the British animal chemistry comes from that of Lavoisier, with his emphasis on analysis of substances into their smallest pieces. Historians like Frederick Lawrence Holmes emphasized that the analysis of the chemistry of life was central to Lavoisier's work, especially toward the end of his life [5]. Toward the end of Lavoisier's life, he was trying to understand how animal respire and how heat affected the processes of life. Lavoisier was trying to determine what animal heat was, and thought that it had to do with an inner burning process, like respiration [6]. Before he could complete his research, he was imprisoned as a former tax farmer and guillotined [7]. In many ways the historiography of animal chemistry is sandwiched between two Holmes books, as the next book that Holmes wrote about animal chemistry is about the work of Claude Bernard, who was working in the middle of the nineteenth century [8].

Claude Bernard, a French researcher, investigated the major organs of the body related to the chemistry of life, such as the liver, as well as the pancreas; he also studied diseases processes related to diabetes. Animal chemistry made a major shift in the middle of the nineteenth century with Bernard's work. Holmes historicized its importance [9]: "the rapid development of methods for identifying and analyzing organic compounds had made it essential to investigate such vital phenomena as respiration, digestion, and nutrition in the light of the new chemical knowledge." This major nineteenth century shift intensified in the early twentieth century and post-war period, as in the work of Robert E. Kohler [10].

This chapter is a good example of how Lavoisier, and other chemists investigating the process of life, turned toward the analysis of the body's stones and how essentially the body's own chemicals contributed to blockages, which were thought to cause disease. The chemistry at the time was not definitive and often times medical providers and chemists put their ideas out in print, or even their own experiences, in order to make a dialogue about how these experiences might contribute to theories or ideas about how to understand concretions. Humoral ideas were still in the conversation, though they were subtle in practice, and perhaps in some cases unrealized by the investigators.

2.1 What Was Animal Chemistry During the Early Nineteenth Century?

William Brooks Johnson, from Coxbench, England, was a complicated philosopher: including ardent supporter of the French Revolution, a participating member in intellectuals circles in England, and a chemist. He published *A History of the Progress and Present State of Animal Chemistry* in 1803 [11]. He was very active in intellectual circles in Europe, such as the Derby Philosophical Society, while he was a student of medicine and chemistry at Cambridge [11]. Johnson was a curious student and physician, as he traveled to South America, where he collected botanical specimen and ultimately constructed a herbarium from the specimens he collected in there. He was in the same society as Erasmus Darwin, who would later make him a member of the Linnaean Society [12–15]. Johnson would go on to publish *Botanic Commentaries* with the help of another botanist Johnathan Stokes. Johnson died in 1830, and *Botanical Commentaries* is dedicated to Johnson's memory.

Johnson's role in the revolution was quite controversial. Johnson, like other characters in this book such as Thomas Cooper, were English supporters of the French Revolution. Johnson, along with colleagues from Derby Philosophical

Society, attended the National Convention in Paris during the early days of the Revolution. Johnson and another member of the Derby Philosophical Society, Henry "Redhead" Yorke, went to the Convention with the task of bringing ideas from the Revolution to plant them in England. However, Johnson and Yorke did not agree with the radical changes of the Revolution, especially its violent turn, and the ideas of Jacobins [12–15]. As the Revolution became more and more extreme, the two men tried to distance themselves from it. Johnson was able to meet Thomas Paine in Paris, but became alarmed and frightened that the people of Paris would follow radical calls from Jean-Paul Marat, who was commanding that all foreign nationals in Paris be murdered. The intensity of Johnson's fear was so severe he tried to commit suicide by stabbing himself with a knife. Johnson was lucky and his life was saved when a doctor was called for his aid. Johnson was later well enough to testify at Marat's trial when Marat was arrested in another move by an increasingly radical government in France. Johnson, at trial, stated that the fear had made him temporarily mentally ill and caused his suicide attempt. Marat was later murdered and Johnson then ran to Switzerland, as they assumed that they were next on the arrest list of Maximilien Robespierre [12–15]. The anxiety of this time period stayed with Johnson.

When Johnson composed the beginning preface of *History of the Progress and Present State of Animal Chemistry,* he emphasized the idea of balance when thinking of one's health, but refused to speculate on the primary cause of disease. Johnson speculated nonetheless [16],

> That all organized bodies are composed of ultimate or remote parts, will not be denied, but there seems to be a prevalent idea, that it is necessary for these component parts to be in certain proportions or peculiar combinations in order to establish the equilibrium of health; and that when these are destroyed, disease is the consequence, which cannot be cured until the organic machine received those parts which are deficient, or gives up those parts which are in excess, so that the proper proportions be again effected, and the equilibrium of health restored.

He emphasized that balance produces a happy and disease-free body, and one that is out of balance is one suffering from disease. Chemistry opened a vista on the disease process links between the mind and the body, as he was optimistic after chemistry had shown secrets about the nature of water [17]: "… who could have believed, thirty years ago, that either fixed air or water was capable of decomposition, that the diamond was nothing but charcoal, and that flesh might be converted into spermaceti? Yet this science is only in its infancy, whilst its progress is deemed to be external." The link regarding the mind and the body in terms of disease is something that was discussed in nineteenth century medical philosophy [18].

Johnson calls those interested in medicine "chemical physiologists," who examine medical ideas through experiments. He was optimistic that chemistry revealed how physicians could balance the body and prevent disease processes. The new chemistry, Johnson promoted, would not allow current researchers to fall into "quicksands" of the past [17]. Johnson was confident in this new chemistry because it had had notable success, such as explaining bodily process while digestion, sweat, or how oxygen works on the vital forces of the body. He hoped that this new bodily chemistry would reveal laws and that philosophers would understand the body and mind. Animal chemists broke the body down into eleven essential parts or substances which included the external parts, poisons, fluids, solids, aromatics, oils, excrement, hard parts, acids, coloring matter, and concretions [17].

The prime substance of the body, however, were the fluids, as Johnson thought that all of the organs of the body originally developed through the secretions, as even the brain was thought to develop from fat. The blood was also important as a principle, but not yet understood, vital fluid. The body was constantly removing urine and sweat, as well as feces from the body, as "useless" fluids were continually removed [19]. It was in the careful chemical analysis of the named eleven parts of the body above that truth could be derived by chemists about the body. However, he does concede that animal chemistry was the weakest part of chemistry at the moment [20].

Concretions that animals produced were of interest to chemists like Johnson because they caused diseases by blocking the body's secretions. These morbid concretions were found in the lungs, bowels, kidneys, and the blood, which were thought to be the "reservoirs" of the body [20]. Johnson was optimistic that examining these concretions would give the chemist insights into the bodily processes that produced them and allow avoidance of disease.

It was in the caverns or passages of the body where concretions formed. Johnson used the calculi that formed in the mouth, the *Calculus salivaris*, as causing swelling and fistulas openings under the tongue [21]. These stones often were thrown out of the fistulas that they created, and were formed from thick saliva, and tarter build up in the parotid area. Other stone of the body, speculated Johnson, were thought to cause arthritis, as stones could appear in the muscles or the bones, as in the case of *Calculus muscularis* [22]. The *Calculus muscularis* was thought to form between bone and muscles, but no calculi had been collected. Other philosophers like William Wollaston described them as much like bone, and could also be found near the heart and veins [23]. Other calculi that formed near particular organs included the *calculus pancreaticus*, and the *calculus pulmonarius*. Calculi that formed near the pancreas resembled salivary stones. Calculi forming the lungs were often expelled through cough and often occurred in the elderly, and often were found in older persons, and the stones were thought by philosophers to cause arthritis. Stones in the uterus, *Calculus uterinus*, were tricky for chemists, as it was hard to tell the stones apart from masses that appeared in the uterus, that he equated with "false conceptions" [24].

2.2 Major Chemists Investigate Stones

William Hyde Wollaston was a major chemist in British history [25]. He was from Norfolk and was educated at Cambridge, where he studied medicine, graduating in 1788, and then completed his doctorate at Cambridge in 1793. In between his medical studies, he practiced in London from the time of his first graduation from

Cambridge to 1801. He stopped practicing medicine because of the stress of seeing patients in great pain [26]. However, some scholars would argue that Wollaston left medicine because he did not get an appointment to a prestigious hospital, such as St. George Hospital. But he did receive membership in 1795 to the Royal College of Physicians. Wollaston would also be involved in the Medico-Chirurgical Society, which is discussed later in this book. Gradually, Wollaston shifted his research from medicine to chemistry. He is most well-known for his work on elemental chemistry, such as work on platinum, and was able to produce the metal through cooperation with a partner [25, 27]. He was made a member of the Royal Society and was awarded the Copley Medal, the oldest and most prestigious award of the Royal Society for his work in chemistry in 1802. He later served as secretary and vice-president of the society. Wollaston suffered with blindness later in life, eventually developing a tumor in the brain. After disclosing the secret steps for making platinum in the lab, the results were communicated to the Royal Society, and he received the medal of that same society prior to his 1828 death.

Wollaston's chemical research also included the studying of concretions. Wollaston was able to develop typologies regarding urinary stones. He developed four types of stone classifications in 1797, as well as studied gout stones. He also argued that contrary to previous chemical ideas, gout stones were not principally chalk, but made of animal materials, much like bone or earth based hard substances [28]. Chemists as far back as William Cullen, a surgeon in the eighteenth century, thought that gout stones could also be destroyed with acidic substances, like nitrous acid, but Wollaston disagreed [29]. Wollaston also argued that lithic acid was not the sole producer of gout stones, and chemistry would ultimately reveal a primary acid that was responsible for forming the stones [29]: "Had he undertaken an examination of the substance itself, he would have found that, instead of a mere concrete acid, the gout matter is a neutral compound consisting of lithic acid and mineral alkali; as the following experiments will prove."

In his chemical research, he worked with several gout stones to determined their chemical essence [30]. He was able to transform the stone through distillation, as he added vitriolic acid, which made the stones dissolve into salts and white powders, which showed him that the stones were made up of some type of animal matter, as there was also oil and an odorous "alkaline liquor" [30]. Wollaston also thought that he could reconstruct a gout stone by combining those same elements. Wollaston proposed that those other gout stones made of chalk were really comprised of the lithic acid and mineral alkali that he had found by distillation. Wollaston had larger goals for studying gout stones, it might contribute to some type of treatment for those patients suffering from gout stones. Optimistic about the fruits of such chemical analysis, Wollaston hoped that [31],

> The knowledge of this compound (lithiated soda) may lead to a further trial of the alkalies, which have been observed by Dr. Cullen to be apparently efficacious in preventing the returns of this disease ...; and many induce us, when correcting the acidity to which gouty persons are frequently subject to employ the fixed alkalies, which are neither of them capable of dissolving gout matter, in preference to earths (termed absorbent) which can have no such beneficial effect.

Wollaston made similar claims about the potential for acids to treat other stones of the body, such as urinary stones. He proposed that [32],

> When a calculus is discovered, by evacuations, to be of the fusible kind, we seem to be allowed a more favourable prospect in our attempts to relieve: for here any acid that is carried to the bladder will act upon the triple crystals, and most acids will also dissolve the phosophated lime; while alkalies, on the contrary, would rather have a tendency to add to the disease.

2.3 The Chemistry of Gout: Physicians and Surgeons

Physicians like Charles Scudamore took inspiration from Wollaston's work and started analyzing urine and gravel found in the urine from patients suffering from gout. Scudamore published his work in *A Treatise on the Nature and Cure of Gout and Rheumatism*, with a third edition published in 1819 [33]. Scoramore shared the same optimism about chemistry as a tool to improve patient treatment that Wollaston shared in his earlier writings. Scudamore was convinced that chemistry yielded a type of truth in science [34]: "… in formation of every science, the collision of ideas, and comparative observations, invariably tend to elicit information and establish truth."

Scudamore examined the urine of a patient suffering from paroxysms [35]. He observed the patient and examined the patient's lifestyle and urine. Some physicians thought that diet effected the disease process of gout. The urine was looked at by physicians and chemist to indicate the disease of gout. If the patient voided urine that was pink, or had the character of "brick-dust sediment," then it meant that the body was either suffering from gout, or developing gout [36]. Reading urine was an important in understanding the patient's suffering of gout [37],

> Whether gout be present, or whether a patient, free from the tendency to that disease, be affected with the disposition to secrete urine of this character, we may consider not only that the digestive function are irregular, but that the different secreting organs are much influenced by a nervous state of the constitution.

Urine was important because it yielded other aspects of information, such as gravel and crystals indicating disease processes. Gout was also thought to signal itself early through urine, such as the patient having very acidic or what Scudamore calls "animalised urine" [36]. The body conveys irritation that often results in a stone or small amounts of gravel in the body's passages. The presence of gravel is indicated in the color of the urine, especially as the severity of gout picks up in the patient. And gravel and the other stoney materials produced in the urine can yield information that one day might help improve patient treatments [38]: "In short, so far as my experience has gone, the sediments in question present as much diversity, as the formed calculi of the bladder, and require the same pathological reasoning, and medical treatment." Stones produced in the urine in cases of gout inspired chemists and physicians to continue research into stone formation in order to improve treatment and understand why stones form.

Research into gout stones was not limited to chemists and physicians. Surgeons contributed to information about the chemistry of gout stones. James Moore, born in 1762, was a surgeon in England, who served in the Second Regiment of Life Guards, wrote about the relationship between the body's fluids and gout stones. Moore served in the American Revolution, who studied at the University of Edinburgh as well as St. George's Hospital, and was a contemporary of Edward Jenner. Moore passed away in 1860 [39]. Moore was also a member of the Medio-Chirurgical Society, and he published his paper "On Gouty Concretions; or Chalk Stones," that discussed the formation of different types of chalk stones that formed in the passages of the body, especially in the cases of gout [40].

In his paper, Moore noted that physicians struggled treat gout, but the resulting chalk stones. But he cautioned physicians treating gout stones as essentially beyond their abilities [40]: "Physicians are seldom competent to enter minutely into the treatment of those tumors and ulcers which are occasioned by chalk stones; and this subject has been unaccountably neglected by surgeons, in whose province they lie." He wanted to get the society motivated to treat stones and offer some solutions. Moore tried to lay out the chemical differences between "concretions" and what he considered "chalk stones," as he counted them as chalk stones because they were made up of lithic acid and not any lime-based substances, but conceded that this chemical knowledge was limited to knowing these differences and focused on treatment interventions meant for the "practical surgeon." Nonetheless Moore demonstrated a chemical literary that was commanding regarding understanding gout stones.

Moore was interested in the types of fluids that were produced in gout disease. White liquid was filling the cavities of those suffering from gout. The fluid spreads throughout the body as the patient continued to suffer from gout. He thought that gout originated in the hands and feet, and he names them the seat of the disease. The seat of disease is a humoral term, which was the area where diseases started [41]. Inflammation was observed in the hands and the feat, where stones appeared in cysts. Irritation continued to manifest in the skin. Other philosophers thought that there was a watery blood that conveyed irritation throughout the body [42]. The white, or "chalky liquid" traveled from the fingers and if the body cannot absorb enough of the fluid, it forms into masses and likely spreads the disease, causing symptoms like paroxysms [42]. Pain spreads in the joint, and much like the Paracelsian idea of an internal alchemist being overwhelmed and not able to sort the substances coming into the body, the body becomes overwhelmed by the fluid and produces disease [43, 44].

The resulting concretions from gout caused the patient pain and often led to poor healing and ulcers. The white liquid could also leak out the body in some cases [45]. And the stones related to gout did real damage to patients [45]:

> when a violent fit of the gout attacks a chalky tumor, the appearance is frequently very alarming, the new paroxysm being accompanied with a fresh serous and chalky effusion,

> which added to the old deposit of chalk, occasions a prodigious swelling, the cutis when distended to the utmost opens; yet sometimes the cuticle remains entire. The chalky or serous liquid may then be seen through the semitransparent epidermis. The surrounding integuments appear of a deep red, or a purple hue, threatening mortification; while the pain is excruciating.

Moore also thought that once the white liquid was excised from the body the patient recovered. The levels of the white chalky substance in the body also might indicate the severity of disease.

As a surgeon, he was also focused on tangible treatments for gout. He did not think that any current treatment really worked well. Perhaps puncturing the body with small wholes might be therapeutic, which again, was thought to relieve the patient from suffering from gout. Getting the inflammatory process under control would also allow the surgeon more opportunity to act. Substances like *Kali purum* or *argentum nitratum* worked on the stones that remained in the body [46]. It was unclear about how much diet and exercise could help patients suffering from gout, and this also depended on the severity of the case.

2.4 Systematic Research into Concretions: Chemists Examine Diabetes, Blood, and Other Fluid-Based Diseases

Chemists were excited to contribute to research regarding fluid-based disease that healthcare providers speculated were caused by concretions. Continuing the work of Wollaston was Thomas Thomson. Thomson was born in 1773 in Perthshire, Scotland, and he was educated at St. Andrews University [47, 48]. Thomson studied medicine there where he received a doctorate in medicine, wrote and edited the *Encyclopedia Britannica*, and also worked as tutor in chemistry to private students from 1800–1811. He was a contemporary and friend of philosophers such as James Mill and Jeremy Bentham, and then served as editor of *Annals of Philosophy* and *Literary Journal.* He was married to Agnes Colquhoun, and his children Thomas and Margret Agnes became naturalists in their own right. His teaching book series, *System of Chemistry,* went through six editions. He was also active in intellectual circles in Britain, such as the Medico-Chirurgical Society and the Royal Society of Edinburgh. Thomson died in 1852.

By the turn of the nineteenth century, Thomson's *System of Chemistry* was widely popular and in its third printed edition, and was an important text for students learning chemistry and those philosophers engaging in serious chemical research. The comprehensive book of chemistry covered most every chemical subject, from plants to morbid concretions and calculi found in the body. Thomson defined morbid concretion and calculi. There were "Hard Substances" were found in animal bodies, especially the cavities of the body. He promoted that fluids in the body explain life and death, as well as disease processes. Understanding the fluids of the body was crucial to improving medicine [49]:

> Their formation [morbid concretions] is an irregularity in the animal economy, and is likely therefore to throw light upon its functions; for we succeed best in detecting the secrets of Nature when she deviates from that regularity which she usually follows. They often produce the most excruciating disease; and the sufferings of the patient can only be effectually relieved removing their causes.

Thomson organized the types of concretions of animals and people into five main groups: urinary calculi, gouty calculi, intestinal concretions, officiations, and biliary calculi [50]. He also categorized the fluids of the body to include: pus, dropsy liquor, and blister liquor [50]. Chemists had researched the fluids collected from patients suffering from various diseases in hospitals in order to get more information about how those liquids play a role in disease.

Bone like concretions were categorized as ossifications, and they occurred in different parts of the body. Some ossifications include those that occur in the brain, pancreas, salivary, and in the lungs. Ossification in the lungs were thought to irritate people into coughing, or cause other disease like consumption. Thomson noted that these stones were found in dissections where the patient was suffering from pulmonary diseases, as surgeons found and analyze these stones. Stones also occurred in the liver and the prostate. These stones could also be found inside muscles and also in blood vessels, but were often overlooked because they resembled bones.

Stones also occurred in the intestines and in the digestive organs of animals. These stones occurred rarely in humans, though a dissertation was constructed about such stones that will be discussed in the seventh chapter of this book. Folk medicine privilege stones from the stomach and intestines of animals as they were believed to having healing properties. These stones are called "bezoars," which are also discussed in the Harry Potter book series. Chemists would also study bezoars from animals to generalize about intestinal stones. Stones also formed in the bile ducts and gallbladder as well. These stones could cause disease by blocking the intestines and causing the patient to suffer from jaundice, which is a yellowing of the eyes and skin.

Thomson shared chemical research about urinary stones and emphasized their importance in medical chemistry [51]: "Chemistry had no sooner made its way into medicine than it began to exercise its ingenuity upon the urinary calculi; and various theories were given to their nature and origin." He also discussed the importance of gout stones and their problematic nature for the joints of patients. Thomson, as an authority on the teaching of chemistry, included the chemistry of bodily stones along with other important chemical topics to be learned by the student. Medical chemistry was becoming an important and interesting tool in the practice of physicians and surgeons. Chemists, physicians, and surgeons would extend their analysis of disease based in the body's fluids to other disease beyond blockages and gout, and this would include diabetes.

2.5 Medicine, Chemistry, and Diabetes

Diabetes was a disease that was not fully defined or explained during the early nineteenth century. Some healthcare providers thought that it was a problem caused by concretions blocking the fluids of the body, while others thought that it was a disease exclusively related to the blood or the urine. Diabetes is another case study in regards to diseases that were thought by some chemists, physicians, and surgeons to be related to fluid blockages of the body, often caused by concretions.

In Ree's *Cyclopedia* the nature of diabetes is discussed in vague terms, noting that it was "… a profuse excretion of urine" [52]. Diabetes also had several names in the *Cyclopedia:* fluxus urina, diarraea ad urinas, diarrahe ad matula, and other names. The entry also included the defining characteristics of the disease as hunger, thirst, tight skin, and general weakness. And chemists were interested in the disease as it involved researching the fluids of the body. One example of such a chemist was William Prout, who saw a disease like diabetes as having several related symptoms. He remarked that [53], "the term *diabetes*, implying simply an increased flow of urine, is applicable to any disease which that symptom is present in a remarkable degree." Though there were a lot of discussions and attempts to define diabetes, most of them involved an over abundance of urine that the patient needed to remove from their body, and this was often accompanied by pain in the urinary organs. Also important to diagnosing the condition, according to Prout, is detecting sugar in the urine. And this takes a certain degree of uroscopy, as Prout emphasizing the detecting of sugar in the urine involves identifying color, smell, and other physical descriptive information. The urine often had a "pale straw or greenish colour" or a milk like appearance, or in some cases there is blood that appears in the urine. The bowels are also irritated and other parts of the body display inflammation, such as passing gas or irritation in the gums, bowel problems chills, heart problems, or a "hay" characteristics of breath [54].

However, Prout refuses to engage in general speculation about the reasons that diabetes occurs in patients other than sugar appeared in the urine and there was an increase of urine [55]. A healthy body was therefore a body that expresses urine, but not too much. He then changed his discussion to the treatment of diabetes and noted that there were a lot of rumors about the condition [55]: "With respect to the *treatment* of diabetes, this has been as various as the opinion respecting its nature; and, perhaps there is no disease in which so much mischief has been done upon false principle and by random experiment, as in this." Prout's treatment recommendations involved getting control of the patient's urine flow, perhaps calming their nerves, but also performing bloodletting [55]. He also advocated cleaning of the intestines, as well as drugs that purged the fluids of the body, as well as opium, baths, warm clothes, and the moderation of a curative regime. Also food and drink should also be restricted [55].

Physicians interested in chemistry also studied diabetes, specifically *diabetes mellitus*. John Bostock, however, wrote about the less covered *diabetes insipidus*. Bostock shared a paper on the topic in 1812 [56]. Bostock won a medal from the Medico-Chirurgical Society from his 1804 work. Bostock was an active and curious physician. He worked in fever hospitals, constructed a botanical garden, Geology Society, pursued zoology, and helped create intellectual culture by helping to found the Royal Philosophical and Literary Society. He also helped to found the Medico-Chirurgical Society. He was treasurer of that society [56].

Bostock discussed the rare diabetes insipidus in his paper read in 1804 entitled "Two Cases of Diabetes, with Observations on the Different States of this Disease" [57]. The difference between *diabetes mellitus* and *diabetes insipidus* was in the nineteenth century how much matter was expressed by the patient, usually this was in regards to their urine. According the National Institute of Diabetes and Digestive and Kidney Disorders at the National Institute of Health, *diabetes mellitus* is associated with high blood sugar but the patient's kidneys are able to function properly and remove sugar from the blood by expressing it through urine. *Diabetes Insipidus* is when the sugar in the patient's blood is normal but the kidneys cannot remove enough sugar from the body [58].

Bostock's 1804 paper conveys the history of a W. H., who was forty years old, and Bostock shares no further biographical information. W. H.'s urine was put to chemical analysis by Bostock in order to attempt to make sense of the disease the patient was suffering from. Bostock felt that he needed to share the patient's background and contextual clues that might contribute to the disease. The patient had a lot of experience in cold and wet weather, as he was a teamster that drove horses as his profession. The patient self-reported that he had a bad diet, and he thought that diet contributed to his illness. W. H. thought that he had been ill for about twelve years, with complains of pain in his loins, throat pain, excessive urine, and feeling weak. He had engaged in folk medicine, as he went to a country doctor who had tasted his urine and informed him that it tasted sweet, and the folk medicine practitioner gave him vegetable cures, such as the *centaura nigra*, which has been tested even in modern times as a potential medicine for diabetes [59]. The patient reported that the *centaura nigra* helped him with the excessive amounts of urine but the illness had recently returned. W. H. complained of chills, pains in his stomach, lose teeth, camps in his body, including his hands [60]. Bostock was concerned that the patient had no night sweats and had pain in his penis after urinating, with difficulty breathing [60]. Upon inspection of the patient's urine, Bostock observed that the patient expelled urine that had a "pale" as well as a "straw" character to its odorless appearance; eventually the urine developed a crust to it [61].

Bostock applied reagents to the urine sample, such as water and alcohol, which showed the samples was characteristically different that regular urine. He noted that urine from patients suffering from diabetes had urine that had a heavier weight, especially in the case of *diabetes mellitus*. But in *diabetes insipidus*, the urine sample weighed much less than normal urine. Bostock's chemical tests did not reveal any sugar. There were irregular amounts of other chemicals in the urine, such as salt, phosphorus, and urea.

Constitutionally, there were differences between the two cases of diabetes. In *diabetes insipidus* the patient lost energy and simply wasted away. Often there was no sugar detected in the urine of *diabetes insipidus*. Urine output continued to increase while the patient continued to get sicker [62], "There was in this case less emaciation than is usually observed, which may be ascribed partly to the chronic state of the disease, but more to the small quantity of solid matter, which was carried out of the system by the urinary discharge." However, in the case of *diabetes mellitus* Bostock is silent in its comparison.

In the 1812 paper, Bostock further comments on the rarer cases of *diabetes insipidus*, proposing that *diabetes insipidus* results from cases of *diabetes mellitus*. In this paper, Bostock tried to put a lot of emphasis in analyzing the urine, like he performed on the blood in the previous paper [63]. He cited the work of a friend and peer, Alexander Marcet, who was a chemist interested in studying stones, as well as how the fluids of the body are affected by diabetes. Bostock cited the work of Marcet in studying bloodwork of patients suffering diabetes, praising Marcet's work [64]: "in this respect the urine seems to be analogous to the albuminous fluids of the body: for in the elaborate experiments of Dr. Marcet, which have been related to the society, the salts were always found nearly in the same quantity, however, various was the proportion of the animal matter ...." Bostock, like Wollaston mentioned above, was able to reconstitute diseases fluids, as he was able to take diabetic urine and transform it into something that looked like blood, but did not continue to discuss the findings in the paper. But blood and urine seemed to signal to chemists and physicians that these fluids contributed to disease processes and needed to be examined further.

Chemists like Bostock saw the potential for chemical analysis of the fluids of the body and he was not alone. Jon Jacob Berzelius also thought that the secretions of the body held answers to disease [65]. The Swedish chemists often published his ideas in English, and it was also translated, like the work of Lavoisier, by chemists for other chemists. He argued that the animal fluids held answers for chemists understanding disease [66]: "there exists no problem in chemistry more difficult to solve, than that of the secretion of animal fluids." The fluids of the body were a frontier for chemists [67]: "The circulating fluid carried to the organized laboratory which nature employs, no foreign ingredient is added, no chemical regent is interposed, and yet the fluid which flows from these organs has acquired chemical properties, which render it decidedly different from the common circulating masses." The fluids of the body were organized into two types: *secretions* in the first group and *excretions* in the second group. Berzelius defined the secretions as liquids produced in the animal body like milk or urine, or even sweat. These fluids were defined as mostly acidic. These secretions are ultimately all derived from the blood. It is also from the blood that other important liquids for the body are produced, such as eye humor, mucus of all type, bile, and saliva [67]. The mucus that Berzelius discussed also included those that were in urinary, nose, and gall passageways of the body [67]. Berzelius highlighted the important work that he felt that Bostock was doing in regards to the blood, but also admitted that the blood was very difficult for chemists to study. Chemists during the time had reduced the blood into its smallest parts, continuing the reductionist maxim of Lavoisier. They broke it down into its smallest parts [68]: "fibrin, albumen, gelatin, red globules, soda, some neutral and earthy salts, a small portion of sulphur, and a peculiar photosphere of iron, all held in solution by a large quantity of water." Also, Marcet provided samples of blood to Berzelius in the same way that he receives urinary stones from Astley Cooper in a later chapter of this book; there was cooperation between physicians, chemists, and surgeons regarding the fluids of the body and stones. The blood was broken down into its smallest parts the same way that urinary and other concretions of the body would be by stones. If chemists could find a substance smallest parts, then they thought they could understand disease.

In regards to the role of sugar in the urine or blood in diabetes was a mystery that was discussed in chemical communities. These discussions often took place in the papers of intellectual societies and in their journals. For instance, in 1811, the chemist William Hyde Wollaston wrote a response to Marcet in regards to the role of sugar in the bodily fluids of diabetics [69]. Wollaston felt like the question as to nature of sugar in the diabetic blood was answered in the work of John Rollo [70]. Rollo wrote *An Account of Two Cases of the Diabetes Mellitus* which was an intellectual accounting of the knowledge about diabetes during the late eighteenth century that he revised throughout his life [71]. Rollo cited the work of Alexander Marcet. In Wollaston's response to Marcet, he did express some hesitation about the limits of chemistry to studying diabetes [70]. Wollaston, who was no longer practicing medicine at the time of writing the letter, felt that physiology might hold some answers to understanding diabetes [69]:

> … since so much of this subject as is strictly physiological, relating to the natural course of circulating fluids, and more specially so much of the investigations as is conducted by chemical means, is within the range of those pursuits which are generally interesting to the Royal Society, I will endeavor to give you as a distinct an account as I am able of the progress of my own experiments, requesting that you will in return state, more fully than you have hitherto done, the result of the further step in the inquiry which you took at my suggestion, and if it agreeable to you, we will without delay make a joint communication of our researches to the Society.

But Wollaston decided to include chemical information that he discovered, as well as the work of Marcet.

Wollaston noted that Rollo, a surgeon, and worked with another surgeon had engaged in the chemical analysis of testing the urine of diabetic patients. However, Rollo had never tasted the blood to determine if it had a sweet taste. Wollaston noted that it was hard for Rollo and his fellow surgeon, Cruickshank, to determine its taste because they only had a small amount of blood to work with [72]. Cruickshank did put the small amount of blood under chemical analysis and compared it to normal blood. Wollaston tried to repeat the experiments that Cruickshank performed on the blood and compared the sample to normal blood. Cruickshank had found that the blood sample contained sugar. The urine that Wollaston tested from diabetics came with information that Wollaston knew that they had eaten sugar, and was able to isolate the sugar in the urine [73]. But Wollaston noted that he could not taste sugar in the patient's urine, but did taste the sugar in the patient's blood sample. Therefore, from the chemical work that Wollaston performed, he argued that because he could not detect any sugar in the blood chemically, and he could find sugar chemically in the urine, the kidneys or intestines must be to blame for the urine entering into the urine. Wollaston summed up his argument [74]:

> In order to account for the presence of sugar in the urine, we must consequently either suppose a power in the kidneys of forming this new product by section, which does not seem to accord with the proper office of that organ; or, if we suppose the sugar to be formed in the stomach by a process of imperfect assimilation, we must then admit the existence of some channel of conveyance from the stomach to the bladder, without passing through the general system of blood-vessels.

Other ideas about the cause of diabetes could be related to stones in the body affecting the fluids of the body.

2.6 Concretions and Diabetes

Some surgeons speculated that concretions caused diabetes. An example of surgeon who engaged in this speculation was Thomas Cawley. Cawley served the British Army in its colonial occupation of the island of Jamaica. While serving as an army surgeon, Cawley published an account of diabetes that was related to concretions in 1788 [75, 76] He treated a patient by the name of Allen Holford, esq., and his write up of the case takes place from 1787 December through June. Cawley noted that Holford was suffering from diabetes and had been suffering, as he had transformed from a well man in his early thirties to a very weakened version of himself. Holford was expelling urine that was sweet, and when it evaporated, there was a sugary substance that remained [77]. Cawley tried to ease the patient's pain with medicines but Holford died, and the medicines that were administered prior to death were causing the patient to suffer more: thirst, hemorrhoids, and dry tongue, mouth, and skin [78]. The patient suffered extreme thirst that could never be satisfied and eventually lost his mind. Cawley decided to dissect Holford and write up the results of the dissection.

Cawley noted that Holford had normal kidneys but a damaged liver. The liver lacked the tumors that other medical providers thought were present in cases of diabetes [79]. Holford also had a normal gallbladder but an abnormal pancreas. He noted that the pancreas was full of concretions [80]: "The pancreas was full of calculi, which were firmly impacted in its substances." Cawley wrote about the size of the concretions, as he noted that they were small and white, much like "mulberry stones." He analogized them to those that appeared in the glands that produced saliva. He noted that the pancreas was also damaged. He thought that there was a link between the presence of concretions and diabetes.

2.7 Conclusion

Returning to the work of John Bostock, recalling the story about the woman suffering from uterine concretions. Bostock wanted to understand how the body's fluids played a role in creating blockages like those uterine concretions. They wanted to end patients suffering from uterine concretions and other concretions of the body, such as urinary stones. Medical providers interested in chemists like Bostock wanted to make comparisons between the "tumors" that were described in his summary case study of a woman suffering from uterine concretions and those that occur in other patients, such as urinary stones.

Physicians, chemists, and surgeons put concretions to the scrutiny of chemical analysis in order to gather information that they hoped would improve treatment. They also did comparative work, such as that Bostock engaged in where he compared the uterine concretions with that of the salivary gland. Bostock's work signaled that perhaps stones produced in the body might be related or caused by similar processes of the body's fluids. He noted a similar case where he found that the stones chemically [81], "I had the opportunity, many years ago, of examining a salivary calculus, which I found to be composed of phosphate of lime, with a little animal matter, and more lately I found nearly the same constituents in a pulmonary calculus, and in two concretions from the ovaries." Bostock himself made vaguely humoral sentiments in work of more general readership, such as *The Cyclopedia of Practical Medicine* [82]:

> And as we must take the fluids, and the changes proper to the fluids into account in all attempts at explanation of physiological phenomena, so we must be prepared to admit Humoral Pathology as essential to the explanation of all the more important phenomena of disease; but this must be a pathology founded on observed changes, not simply of the chemical of chemical condition, but of the strictly vital properties of the fluids, and especially of the blood.

The fruitfulness of chemical analysis of stones not only inspired an interest in comparing concretions of the body but renewed interest in fluid-based conception of disease. The next chapter on Benjamin Rush explores the interests of physicians applying chemistry to understand how the fluids of the body play a role in diseases processes and the importance of humoral pathology in medical chemistry by some physicians in the early nineteenth century. Some physicians from American to Europe saw all this chemical work as indicative of the return of the old ways of medicine. For instance, the *Medical Times* ran excerpts and summaries prompting [83], "… that humoral pathology, purged of its errors, will once more take a prominent place among medical theories."

References

1. Bostock J (1835) On the chemical constitution of calcareous tumors of the uterus, and other marts. Medico-Chiru Trans 19:81–93
2. Golinski J (1999) Science as public culture: chemistry and enlightenment in Britain, 1760–1820. Cambridge University Press, Cambridge
3. Marcet A (1813) A chemical account of the various dropsical fluids with remarks on the nature of the alkaline matter contained in these fluids, and on the serum of the blood. Medico-Chiru Trans 2:342–384
4. Thomson T (1804) A system of chemistry in four volumes. Bell & Bradfute, and Balfour, Edinburgh
5. Holmes (1987) Lavoisier and the chemistry of life. University of Wisconsin University Press, Madison
6. Poirier J-P (1993) Lavoisier: chemist, biologist, economist (trans: Balinski R). University of Pennsylvania Press, Philadelphia
7. Donovan A (1993) Antonine Lavoisier: science, administration and revolution. University Press, Cambridge, Cambridge
8. Homes FL (1974) Claude Bernard and animal chemistry. Harvard University Press, Cambridge
9. Holmes FL (1974) Claude Bernard and animal chemistry. Harvard University Press, Cambridge, p 3
10. Kohler RE (1982) From medical chemistry to biochemistry: the making of a biomedical discipline. Cambridge University Press, Cambridge
11. Brooks WJ (1803) History of the progress and present state of animal chemistry. J. Johnson, London
12. Lee S (1909) Dictionary of national biography. MacMillan Company, London
13. Robinson E (1953) The derby philosophical society. Ann Sci 9:359–367
14. Sturges RP (1978) The membership of the derby philosophical society. Midl Hist 4:212–229
15. Elliot PA (2009) The derby philosophers: science in British Urban society, 1700–1850. Manchester University Press, Manchester
16. Brooks WJ (1803) History of the progress and present state of animal chemistry, vol 1. J. Johnson, London, p iii
17. Brooks WJ (1803) History of the progress and present state of animal chemistry, vol 1. J. Johnson, London, p iv
18. Driggers EA (2021) "The voice of the stomach": the mind, hypochondriasis and theories of dyspepsia in the nineteenth century. History Pysc 32:85–99
19. Brooks WJ (1803) History of the progress and present state of animal chemistry, vol 1. J. Johnson, London, p x
20. Brooks WJ (1803) History of the progress and present state of animal chemistry, vol 2. J. Johnson, London, p 2
21. Brooks WJ (1803) History of the progress and present state of animal chemistry, vol 2. J. Johnson., London, p 347
22. Brooks WJ (1803) History of the progress and present state of animal chemistry, vol 2. J. Johnson, London, p 348
23. Brooks WJ (1803) History of the progress and present state of animal chemistry, vol 2. J. Johnson, London, p 349
24. Brooks WJ (1803) History of the progress and present state of animal chemistry, vol, 2. J. Johnson, London, pp 349–350
25. Usselman MC (2015) Pure intelligence: the life of William Hyde Wollaston. University of Chicago Press, Chicago
26. Lee S (1909) Dictionary of national biography. MacMillan Company, London, pp 311–316
27. Hinde PT (1966) William Hyde Wollaston: the man and his "equivalents." J Chem Ed 43:673
28. Wollaston WH (1797) On gouty and urinary concretions. Phil Trans 87:386–400
29. Wollaston WH (1797) On gouty and urinary concretions. Phil Trans 87:386

30. Wollaston WH (1797) On gouty and urinary concretions. Phil Trans 87:387
31. Wollaston WH (1797) On gouty and urinary concretions. Phil Trans 87:389
32. Wollaston WH (1797) On gouty and urinary concretions. Phil Trans 87:399
33. Scudamore C (1819) A treatise on the nature and cure of gout. Charles Scudamore, London
34. Scudamore C (1819) A treatise on the nature and cure of gout. Charles Scudamore, London, pp xxiii
35. Scudamore C (1819) A treatise on the nature and cure of gout. Charles Scudamore, London, p 96
36. Scudamore C (1819) A treatise on the nature and cure of gout. Charles Scudamore, London, p 97
37. Scudamore C (1819) A treatise on the nature and cure of gout. Charles Scudamore, London, p 98
38. Scudamore C (1819) A Treatise on the nature and cure of gout. Charles Scudamore, London, p 101
39. Lee S (1909) Dictionary of national biography. MacMillan Company, London, p 804
40. Moore J (1809) On gouty concretions; or Chalk-Stones. Medico-Chirug Trans 1:112–122
41. Carrick PJ (2001) Medical ethics in the ancient world. Georgetown University Press, Washington, p 35
42. Moore J (1809) On gouty concretions; or Chalk-Stones. Medico-Chirug Trans 1:114–115
43. Moran B (2005) Distilling knowledge: alchemy, chemistry, and the scientific revolution. Harvard University Press, Cambridge
44. Weeks A (1997) Paracelsus: speculative theory and the crisis of the early reformation. State University of New York, Albany
45. Moore J (1809) On gouty concretions; or Chalk-Stones. Medico-Chirug Trans 1:115
46. Moore J (1809) On gouty concretions; or Chalk-Stones. Medico-Chirug Trans 1:121
47. Lee S (1898) Dictionary of national biography. MacMillan Company, London, pp 271–272
48. Partington JR (1949) Thomas Thomson, 1773–1852. Ann Sci 6:115–126
49. Thomson T (1810) A system of chemistry: in five volumes, vol 5. Bell & Bradfute, Edinburgh, p 658
50. Thomson T (1810) A system of chemistry: in five volumes, vol 5. Bell & Bradfute, Edinburgh, pp 658, 690–697
51. Thomson T (1810) A system of chemistry: in five volumes, vol 5. Bell & Bradfute, Edinburgh, p 670
52. Rees A (1819) The cyclopedia: or universal dictionary of the arts, sciences, and literature 11. Longman, Hurst Rees, Orme, & Brown, London
53. Prout W (1848) On the nature of the stomach and renal disease: being an inquiry into the connexion of diabetes, calculus, and other affections of the kidney and bladder, with indigestion. John Churchill Princess Street, London
54. Prout W (1848) On the nature of the stomach and renal disease: being an inquiry into the connexion of diabetes, calculus, and other affections of the kidney and bladder, with indigestion. John Churchill Princess Street, London, pp 59–60, 24
55. Prout W (1848) On the nature of the stomach and renal disease: being an inquiry into the connexion of diabetes, calculus, and other affections of the kidney and bladder, with indigestion. John Churchill Princess Street, London, p 72
56. Pettigrew TJ (1839) Biographical memoirs of the most celebrated physicians, surgeons, etc. Whittaker and Co., London, pp 1–20
57. Bostock J (1804) Two cases of diabetes, with observations on the different states of this disease. Memoir Med Soc London 6:237–265
58. NIH (2022) https://www.niddk.nih.gov/health-information/kidney-disease/diabetes-insipidus#what-is
59. Salachna P, Pietrak A, Lopusiewicz (2021) Antioxidant potential of flower extracts from *Centaurea* spp. Depends on their content of phenolics, flavonoids and free amino acids. Molecules 26:7465

60. Bostock J (1804) Two cases of diabetes, with observations on the different states of this disease. Memoir Med Soc London 6:258–260
61. Bostock J (1804) Two cases of diabetes, with observations on the different states of this disease. Memoir Med Soc London 6:260
62. Bostock J (1804) Two cases of diabetes, with observations on the different states of this disease. Memoir Med Soc London 6:265
63. Bostock J (1812) Observation on diabetes insipidus. Medico-Chirur Trans 3:10–122
64. Bostock J (1812) Observation on diabetes insipidus. Medico-Chirur Trans 3:119–120
65. Jorpes JE (1970) trans. Steele, B Jac. Berzelius: His Life and Work. University of California Press, Berkley
66. Berzelius J (1816) General views on the composition of animal fluids. Medico-Chirur Trans 3:198–289; 234
67. Berzelius J (1816) General views on the composition of animal fluids. Medico-Chirur Trans 3(1816):234
68. Berzelius J (1816) General views on the composition of animal fluids. Medico-Chirur Trans 3(1816):232
69. Wollaston W (1811) V. On the Non-Existence of Sugar in the Blood of Persons Labouring Under Diabetes Mellitus. In a Letter to Alexander Marcet, M. D. F. R. S. from William Hyde Wollaston M. D. Sec. R. S. Phil Trans of Royal Soc Lon 101:96–109
70. Wollaston W (1811) V. on the non-existence of sugar in the blood of persons labouring under diabetes mellitus. In a Letter to Alexander Marcet, M. D. F. R. S. from William Hyde Wollaston M. D. Sec. R. S. Phil Trans Royal Soc Lon 101:96
71. Rollo J (1797) An account of two cases of the diabetes mellitus. Ann Med 2:85–105
72. Wollaston W (1811) V. On the Non-Existence of Sugar in the Blood of Persons Labouring Under Diabetes Mellitus. In a Letter to Alexander Marcet, M. D. F. R. S. from William Hyde Wollaston M. D. Sec. R. S. Phil Trans of Royal Soc Lon 101:97.
73. Wollaston W (1811) V. On the non-existence of sugar in the blood of persons labouring under diabetes mellitus. in a letter to Alexander Marcet, M. D. F. R. S. from William Hyde Wollaston M. D. Sec. R. S. Phil Trans Royal Soc Lon 101:99–101
74. Wollaston W (1811) On the non-existence of sugar in the blood of persons labouring under diabetes mellitus. In a Letter to Alexander Marcet, M. D. F. R. S. from William Hyde Wollaston M D Sec R S Phil Trans Royal Soc Lon 101:102
75. Cawley T (1788) A singular case of diabetes, consisting entirely in the qualities of urine; with an inquiry into the different theories of that disease. Lon Med J l(9):286–308
76. Engelhardt D (1989) Diabetes its medical and cultural history: outlines, texts, bibliography. Springer-Verlag, London, pp 470. This is the original citation for Cawley
77. Cawley T (1788) A singular case of diabetes, consisting entirely in the qualities of urine; with an inquiry into the different theories of that disease. Lon Med J 9:286
78. Cawley T (1788) A singular case of diabetes, consisting entirely in the qualities of urine; with an inquiry into the different theories of that disease. Lon Med J 9:287
79. Winslow CEA (1935) A physician of two centuries ago: Richard mead and his contributions to epidemiology. Bull Hist Med 3:509–544
80. Cawley T (1788) A singular case of diabetes, consisting entirely in the qualities of urine; with an inquiry into the different theories of that disease. Lon Med J 9:289
81. Bostock J (1835) On the chemical constitution of calcareous tumors of the uterus, and other marts. Medico-Chiru Trans 19:82
82. Bostock J (1832) History of Medicine in the cyclopedia of practical medicine. London, Cherwood, Gilbert, Piper, Baldwin, Cradock, Paternoster-Row
83. (1840) The Medical Times. J Eng For Med 2:164

3 Race, Concretions, and Humoral Theory in the World of Benjamin Rush

Benjamin Rush (Fig. 3.1), who lived from 1745 until his death in 1813, was arguably tone of the most well-known American physicians during the founding of America. He taught many students at the University of Pennsylvania that went on to practice medicine in both the northern and the southern regions of the United States. His medical texts went through several editions and served as the basis for a lot of clinical teachers. Rush was also a trained chemistry, who traveled to some of the best centers of chemical research in Scotland. During the American Revolution, Rush was active in the medical department of the Army. He also healed the political and person rift that occurred between Thomas Jefferson and John Adams, as he wrote letters to both men to encourage them to mend their relationship.

Rush believed in the power of bloodletting and saw its needed application to the concretions of the body. In the collections of his medical writings, he discussed treating the concretions of the body, the value of bloodletting is useful [1]:

> 2. It prevents those congestions and effusions which produce apoplexy, palsy, pneumonia notha, calculi in the kidneys and bladder, and chalk-stones in the hands and feet. The gravel and stone are nine times in ten, I believe, the effects of an effusion of lymph or blood from previous morbid action of the kidneys.

Physicians like Rush believed that bloodletting was a useful intervention for a sick patient because it adjusted the patient's bodily fluid and prevented further concretions from forming. Bloodletting, or breathing the vein, was seen by many eighteenth century practitioners as an incredibly useful tool for physicians to restore the body back into its natural balance and allow it to heal itself [2]. Rush saw the value of bloodletting as crucial to treating conditions related to diseases of the stone. For instance, he shared a story about an older man who was ninety-six, who had not been able to pass urine for a long time. Rush saw the man while practicing in Philadelphia and bled him, which cured his condition. He recounted the story to show the value of his intervention [3]: "His only remedy for it was

E. A. Driggers, *Early Nineteenth Century Chemistry and the Analysis of Urinary Stones*, Perspectives on the History of Chemistry,
https://doi.org/10.1007/978-3-031-34973-7_3

Fig. 3.1 Benjamin Rush (1745–1813). Stipple engraving by Edwin, 1813, after T. Sully (welcome collection)

bleeding. He lived free from the gravel and atone, and died, or rather appeared to fall asleep death, from old age."

Rush not only believed in the value of bloodletting to control the fluids of the body, but also thought that the study of chemistry could help to understand the stones of the body. Rush was a curious physician, who was involved in many subjects beyond chemistry and medicine, such as politics and religion [4]. Rush, as a curious person, applied chemistry to many different medical problems, from differences in race to problems related to bodily worms. At the heart of all these projects was the theory that the body's fluids needed to flow, and if they were blocked, problems occurred for the patient. Examining Rush's life in the context of this book shows the association that medico-chemists like Rush had with concretions causing disease, as he believed that African slaves suffered a type of sickness that was reflected in their skin color because of blockages in their internal fluids. Also, Rush was educated in the new chemistry and brought his humoral medical views into his chemical analysis.

Rush was practicing chemistry while humoral pathology was still very much in the minds of chemists and physicians. Historians like John Powers have highlighted the importance of chemistry in the seventeenth century and how it involved aspects of humoral theory [5]. Powers wrote that one of the most important chemists of the time period, Herman Boerhaave, argued that, "the proper flow of the bodily fluids denoted the health and life of the organism. Retarded fluid flow in some part of the body caused illness, and a complete halt in all fluid flow indicated death" [6]. Boerhaave thought that disease came from when the system of the body become irritated. These ideas persisted in the minds of William Cullen, who was Rush's chemistry professor while he was studying in Europe, though these ideas changed between the generation of Boerhaave and Cullen [7]. Rush's chemical work is a hybrid of the two generational ideas [8]. This chapter is an exploration of Rush's work with chemistry and its application to disease, especially those related to the concretions of the body.

3.1 Rush Trains in the Latest Chemical Ideas

Rush had desires of coming back to Philadelphia trained in the latest chemical ideas and hopefully finding a faculty position at the University of Pennsylvania Medical Department (also known as Penn). Rush was born in the Philadelphia area in 1746 [8–11]. After he finished preliminary studies in Maryland, he studied for his undergraduate degree at the College of New Jersey. At first, Rush was going to pursue a career in law, but ultimately decided on medicine. First, he studied as an apprentice under John Redman in Philadelphia but then came under the influence of John Morgan, who was an early figure in chemistry in America [12]. Morgan's interests in chemistry inspired Rush to pursue chemistry as a graduate student, and he left for medical and chemical education at the University of Edinburgh, where he received his M.D. Rush attended the lectures of Joseph Black and other leading Scottish Enlightenment chemists, as well as leading physicians like William Cullen [13–15]. It was Morgan's hope, as well as Rush's, that he would return to a faculty position at the University of Pennsylvania Medical Department, where he would teach chemistry. Rush was able to travel Europe after his graduation and also met up with Benjamin Franklin while he was in Paris.

Rush took up the position of the chair of chemistry at University of Pennsylvania in 1769. Rush was able to publish his chemical syllabus which became an example in America for chemical education [16]. In 1778, Rush was present at the first and second Continental Congresses, where some members began to speak of an independent America from the British Empire. Rush was helpful in the production of gunpowder for the American cause [17]. Rush extended his chemical tutoring not only to men but to women in early America [16]. Rush saw women as important in the education of the future generation of American citizens, but did not see women as true equals [18].

In 1789, Rush took the position of medical theory and practice after Morgan died and mostly focused on teaching the practice of medicine [8]. Rush still continued to teach chemical ideas with his recommendations for medical treatment, but mostly left the practice of teaching chemistry at the University of Pennsylvania.[1] Chemical ideas can be found in his *Sixteen Introductory Lectures to Courses Upon the Institutes and Practice of Medicine, with a Syllabus of the Latter* [19]. He wrote regarding chemistry that [20], "the study of chemistry affords a perpetual source of pleasure, by unfolding the effects of heat and mixture." Rush had his mind in the chemical world, as he mourned the death of Lavoisier [21]: "... wished his execution to be suspended only till he could fill up the measure of his happiness by completing a course of experiments to ascertain a principle in chemistry." He also called for the application of chemical analysis to the problems of nature [21]: "all the sciences bear testimony to the truth of this remark. Earth, air, water, fire, animals, vegetables, and fossils refuse to yield the component parts to any of the means that have been mentioned. They must be tortured by chemical and mechanical agents for this purpose." Rush wanted all of nature and the body to be scrutinized by chemical analysis.

3.2 Chemistry and the Problem of Race: The Legacy of Benjamin Rush

About one hundred years after the work Benjamin Rush, anatomists and physicians were looking for an explanation for blackness in skin. At the University of Virginia, students were examining cadavers of African Americans trying to explain why their skin was not white, which was not surprising, given the white-supremacist nature of the American South [22]. Blackness, or black skin, was thought to be a result of changes in the body's fluids, like the blood. An anatomist at the University, Dr. Floyd wrote that [23]:

> It is natural to suppose that the substance which gives the characteristic black colour to the skin of the negro is probably modified blood pigment, as is pretty generally assumed to be the case in reference to the 'melanin' of the choroid coat of the eye; but this point does not seem to have been made until now the subject of experiment.

Dr. Floyd was proud that his students were some of the first to apply chemistry to understanding racial differences in the skin. However, these ideas in chemistry had been around for 100 years at the time of Floyd publishing his work.

Rush's work regarding the skin of African slaves was originally discussed in the work of Donald J. D'Elia, a twentieth century historian [24, 25]. However, much of the analysis of philosophers of the eighteenth and nineteenth century has not included the chemical research into why there were skin differences [26–28]. There are historians of chemistry that are starting to study the racial aspects of chemical analysis [14, 29, 30]. Benjamin Rush became interested in the work of Samuel Stanhope Smith. Smith lived from 1795 until 1812, and was the president

[1] An excellent biography of Rush's many academic positions at the University of Pennsylvania can be found at https://archives.upenn.edu/exhibits/penn-people/biography/benjamin-rush/.

of the College of New Jersey, Rush's alma matter, in the late eighteenth century.[2] Samuel Stanhope Smith had a similar viewpoint to Rush, Smith believed that all people were humans, and skin differences were the product of external factors, and skin color could be changed by changes in weather or circumstances.[3]

Smith wrote out his theories on race entitled [31–33], "An Essay on the Causes of the Variety of the Complexion and Figure in the Human Species," which then he expanded into a book. Rush published his response to the book, also validating the work of Smith [34]:

> I admit the Doctor's facts, and reasoning as far as he has extended them, in the fullest manner. I shall only add to them a few observations which are intended to prove that the color and figure of that part of fellow creatures who are known by the epithet of negroes, are derived from the modification of the disease, which is known by the name Leprosy.

Rush was an antislavery advocate, but was not an advocate of equality in regards to African slaves and Europeans [35, 36].

Enslaved African skin color appearance, in Rush's medical opinion, was evidence of disease, much like a symptom. Enslaved Africans were sick white people suffering from leprosy. He blamed the disease on the natural characteristics of the land in Africa, the heat was to blame for disease like fevers but also in what he called African slave's uncivilized behavior [37]. The idea of the disease of leprosy turning skin black as part of its disease processes had appeared in the medical literature, as people from the middle east were associated with dark skin and leprosy [38, 39]. In the eighteenth and early nineteenth century medical mind, the skin was reflecting the internal problems of the body. Rush also cited the Bible as a medical authority, as he cited a case where the disease had caused people to experience [39] "preternatural whiteness" or people that he argued were "albinos." Rush thought that based off of reading New World encounters with albino people that perhaps these people were suffering from leprosy. He would extend these ideas to an enslaved African that had several spots of different colors of white and black. Charles Wilson Peale, born in 1741 and living until 1827, had painted a likeness of the man [40]. Peale was a famous painter of the eighteenth century and painted many famous Americans like Washington and Thomas Jefferson.[4]

Leprosy was an internal problem that had outward symptoms and Rush saw it as a problem of overall health. Such a problem of overall health would have been seen by Rush, an adherent to humoral medicine, as a constitutional issue. Something

---

[2] There is an excellent biography of Samuel Stanhope Smith at Princeton's online historical project about slavery at the institution: https://slavery.princeton.edu/stories/samuel-stanhope-smith.

[3] Samuel Stanhope Smith ideas about race had really wide term effects, see scholars like Greene JC (1954) The American Debate on the Negro's Place in Nature, 1780–1815. J Hist Idea 15:384–396 and Juengel S (2001) Countenancing History: Mary Wollstonecraft, Samuel Stanhope Smith, and the Enlightenment Racial Science. ELH 86:897–927.

[4] Peale was an influential painter in the greater Philadelphia area, and his painters can viewed at the old bank building in Philadelphia. One gallery has posted a well written biography, but more books needed to be written on his art: https://www.nga.gov/collection/artist-info.1774.html.

was irritating the nerves or the senses of the body. He referred the reader to a Doctor Benjamin Moseley, who was a physician who argued that African slaves had a different pain toleration than white people [41, 42]:

> … They [Negroes] sleep sound in every disease, nor does any mental disturbance ever keep them awake. They bear surgical operations much better than white people, and what would be cause of insupportable pain to a white man, a negro would almost disregard. I have amputated the legs of many negroes, who have held the upper part of the limb themselves.

In addition to citing their comparative stronger sexual needs, he attributed physical characteristics of large lips and horizontal nose as indicators of leprosy. Hair was a characteristic that Rush was unable to link to disease or the weather of Africa.

Rush thought that he would be criticized about his ideas, as Smith had been criticized for his ideas about African and Europeans being of the same heritage. Rush admitted that there were several unanswered questions related to his ideas, such as how leprosy was passed between generations to continue producing black skin in Africans. He did not think that leprosy was a contagion [43, 44]. Similarly, it was strange to Rush as to why people were not coming down with leprosy and their skin turning colors. There were examples that Rush could cite, however, where people's skin did suddenly change color [44].

In regards to their dark skin, Rush also posited that this did not hurt their overall health. Rush contextualized his idea in regard to the health of enslaved Africans [44], "… negroes are as healthy, and long lived as the white people. Local skin disease seldom affects the general health of the body, or the duration of human life." Also, leprosy was not thought to affect overall mortality [44]. He also advocated that white people should be easier on enslaved Africans because of their sickness and attribute the difference in skin color not to inferiority, but to sickness [45]:

> … That all claims of superiority of the whites over the blacks, on account of their color, are found alike in ignorance and inhumanity. If they color of the negroes be the effect of the disease, instead of inviting us to tryannise over them, it should entitle them to a double portion of our humanity, for disease all over the world has always been the signal for immediate and universal compassion.

Rush was bolder in advocating the importance of medicine on healing Africans [45]: "… let science and humanity combine their efforts, and endeavor to discovery a remedy for it. Nature has lately unfurled a banner upon this subject."

Rush also offered up additional information regarding his ideas about African slaves being sick white people suffering from leprosy. He discussed the example of Henry Moss, who was an African whose skin gradually turned white over a course of five years. Rush also noted that even Moss's hair had changed. A humoral pathological response was offered up to the reader by Rush to explain the striking changes in Moss. Moss's fluids internally caused him to change, specifically in the "rete mucosum," which was the blood under the skin or the layers of skin themselves. It is unclear what the true nineteenth century meaning of the word is, as most dictionaries today discuss the word as meaning the blood vessels or layers of the skin [46].[5] Rush thought that internal changes caused the rete mucosum or what he termed "colouring matter" to change [47]:

> … perhaps of the rete mucosum itself, for pressure and friction, it is well known, aid the absorbing action of the lymphatics in every part of the body. It is from the latter cause, that the palms of the hands of negro woman who spend their lives at a washing tub, are generally as fair as the palms of the hands in laboring white people.

The fluid was either removed from the body or taken into the body, changing the overall color of the skin, and Rush thought that bloodletting or other humoral interventions could facilitate such as removal of fluids that caused the skin to become darker. He advocated humoral interventions stating that [47]:

> Depletion, wether by bleeding, purging, or abstinence has been observed to lessen the black color in negroes. The effects of the above remedies in curing the common leprosy, satisfy me that they might be used with advantage in the state of leprosy which I conceive to exist in the skin of the negroes.

Rush was convinced that he could heal African slaves and increase their health. His ideas about the black body also had larger implications about equality, as he believed that all races shared a common ancestor, and perhaps argued against the system of slavery as a Christian [47]. However, the chemical ideas that Rush was working through about the fluids of the body continued throughout the early nineteenth century.

3.3 Racial Chemistry in Animal Chemistry

Other chemists had performed chemical research regarding race, including William Brooks Johnsons, who was discussed previously in the introductory chapter of this book regarding the nature of animal chemistry. Brooks was interested in cases where people of African descent had their skin change. In his book *History of the Progress and Present State of Animal Chemistry*, he had examined the chemistry of life and continued the project of Lavoisier. Brooks had examined bodily chemical processes like sweating, digestion, and how the bodily uses oxygen [48]. In the second volume of his three-part work, Brooks examines the question regarding light and the animal body, and how it effects the "Colouring Matter of Animals" [49]. Brooks made arguments about how internal changes in the body produced skin differences between those persons of African and European descent, though the concept of race was a very complicated historical idea [50]. He argued that when animals are exposed to light their skin darkens, and he makes the link that

[5] This fits the definition from the *Oxford Annotated Dictionary*, *Meriam Webster*, and *Cambridge Dictionary*.

[51], "Thus in Europeans we find the parts of the body exposed to light, darker coloured than those that are concealed." He also explained how children of people of African descent have dark skinned children [51]: "Thus we are informed that negroe children when first born are white, and do not become black until after having been a few days exposed to the light, and this blackness is found by observation to be considerably accelerated by early exposure to a strong light." Johnson also theorized in the book that perhaps dark skin was somehow related to how oxygen was changing in the body. He thought that perhaps oxygen "Decomposed" in the rete mucosum in people of African descent, and perhaps the intense subjection to light also had an effect on African children [52].

British chemists continued to attempt to explain blackness, as they viewed it as a pathological, and the non-normal state of skin [53]. Thomas Beddoes was a nineteenth century British chemist who experimented on persons of African descent [54]. In 1790, Beddoes ran an experiment where he attempted to change the skin color of a "distressed negro," from black to white [55]. He was going to use "oxygenated marine acid air" in order to complete the experiment, and the subject agreed to undergo the experiment because he as desperate to gain funds for his family. Beddoes subjected the person's fingers into a jar of the acid, but when the African person felt feelings of itchiness that transitioned into pain, Beddoes expressed worry, and had the subject remove his hand. Beddoes recorded that the fingers of the African subject had changed color into a white, as well as hair that had accidently fallen into the jar during the experiment [56]. However, the African subject continued to experience pain and irritation in his hand the next day, and the subject withdrew from further work with Beddoes.

Natural explanations were also offered by chemists that either hot air or extreme light caused skin color to change. With the increase in knowledge from experiments on the skin and the body's fluids, chemists and other turn of the nineteenth century intellectuals thought that skin color could be continually changed by exerting either natural or chemical influences. These ideas would influence chemists from Philadelphia to London, where their work would be published in society journals. Pro-slavery thinkers in America would continue to promote this chemical work to further validate their pro-slavery agenda in the American South. These chemical ideas could be found in political works, such as Thomas Jefferson's eighteenth century *Notes on the State of Virginia.* Jefferson promoted the difference between white and black people by citing differences in the internal fluids of the body with very humoral ideas [57]:

> The first difference between which strikes us is that of colour. Whether the black of the negro resides in the reticular membrane between the skin and scarf-skin, or in the scarf-skin itself; whether ir proceeds from the colour of the blood, the colour of the bile, or from that of some other secretion, the difference is fixed in nature, and is as real as if its seat and cause were better known to us. And is this difference of no importance.

Jefferson continued his discussion about the difference between African slaves and white Virginians, asserting their differences in a humoral framework of science and medicine. He defined their differences with many of the principles from animal chemistry and humoral pathology [58]:

> They secrete less by the kidneys, and more by the glands of the skin, which gives them a very strong and disagreeable odour. This greater degree of transpiration renders them more tolerant of heat, and less so of cold, than the whites. Perhaps too a difference of structure in the pulmonary apparatus, which a late ingenious experimentalist has discovered to be the principal regular of animal heat, may have disabled them from extricating, in the act of inspiration, so much of that fluids from the outer air, or obliged them in expiration, to part with more of it.

Other experimenters and philosophers will be mentioned in this book that framed the difference between black and white people through chemical and humoral ideas.

3.4 Internal Parasites and Chemistry

Rush wrote an account of the success of treating bodily worms by using chemical substances in 1769 for the American Philosophical Society [59]. He transmitted the story of his treatment of a young girl who had had severe fever and other bodily symptoms of inflammation. He did not expect her suffering from an inflammation-like disease as it was not the time of the year that those diseases struck people. Rush was perplexed, as he had no idea what was causing the girl to remain sick. As he did not know what she was suffering from, he bled and purged her system with laxative medications, and then he advised that she bathe her feet in warm water [59]. The purgative medicines produced worms, specifically ascarides worms, which surprised him. These were likely round worms. Round worms have plagued humanity for many years [60]. Rush, as a chemist, turned his attention to vermifuge medications, or those medications that were known to kill worms. He discussed medicines like "Worm Grass" which originated from Jamaica and the "Pink Root" from the Carolina. Medicines from the Caribbean entered the British and later American consciousness because of the British colonizing Jamaica [61].

Stramonium and its potentially helpful and dangerous affects appeared in discussions in the medical literature that Rush would have read about in medical journals.[6] Rush experimented with Stramonium, which was also a powerful drug, and he was able to gather some from the garden.[7] The child had previously been playing near the garden and had been eating the seeds [62]. Immediately, Rush

[6] See Jonasson M, Afshari (2016) Chronicle of Datura Toxicity in the 18th and 19th Century. Asia Pac J Med Tox 5:101–106.

[7] Rush's gardens have reconstructed, though a bit smaller than the original, in Philadelphia: https://www.nps.gov/inde/benjamin-rush-garden.htm.

made the girl vomit and did so for a week. Though the purge did offer some relief, she was not cured, as he noted that of the patient [63]:

> The pupils of her eyes were much dilated and she catched at the bed—cloathes and at everything around her, in the same manner as a person in the last stage of a fever. I was persuaded the oil she had taken, had evacuated all such of the seeds as were in the guts. I began to suspect that her complains were still kept up by a few seeds which still remained in her stomach. I therefore gave her four grains of tart. Emetic, in the manner I formerly mentioned, and had the pleasure to find, that it brought up above 80 of the seeds, and second time it puked her. Finding the stupor and blindness still continue, I repeated thc puke, which brought up, above 20 more. Upon this all her complaints vanished, and in a few days she appeared perfectly well.

It was likely that that the girl had taken in the seeds, but they must have been fried out, and lost their potency, as he had heard of patient consuming that much Stramonium and not fairing well.

One of the important points that Rush made in this article was the idea that fluids of the body can make changes occur on the skin, such as legions, and these legions can be caused by liquids traveling from different parts of the body, such as the stomach or the intestines. The stomach and the intestines signaled the inflammation of the body and indicated those conditions sooner than changes in the blood. Many of these ideas had very humoral ideas, such as the emphasis on reading the fluids of the body to determine the health of the patient [64].

The second point of the article was a general criticism of the humoral approach to medicine. Rush was frustrated that he could not treat the worms of the patient through traditional humoral approaches like purging. Rush lived in the time of heroic medicine, where strong diseases required stronger cures [65]. Much like the problem in the treatment of urinary stones, the chemical treatments for worms in people were strong medications that were often dangerous. But Rush was convinced that there were natural cures out in nature for worms [66]: "… if we may be allowed to reason from analogy, I think we may presume, that there is scarcely a poisonous substance in nature but what has an antidote provided for it." He was convinced that research and experiments would reveal a useful cure. Rush committed to chemical research to find a better remedy for worms.

Rush communicated to other philosophers about his interest in chemicals and substances, such as his letter to Thomas Jefferson. In 1791, Rush wrote to Jefferson about the value of sugar, particularly its ability to rid the body of worms [67]. He promoted the worm killing nature of sugar to Jefferson [68]: "the plentiful use of sugar in diet is one of the best preventives that has ever been discovered of the diseases which are produced by worms." He continued his research into worms with additional articles, such as "Observations Upon Worms in the Alimentary Canal" [69]. He explored the philosophical questions related to worms, such as how they related to the body, or do they play a role in the process of life, or feed on chemical products related to the life processes. Rush questioned in regards to worms [70]: "Do they consume the superfluous aliment which all young animals are disposed to take, before they have been taught, by experience or reason, the bad consequences which arise from it?".

It was unclear what the role of worms were in disease, and some physicians like Rush thought that in order to maintain good health, people even needed a balance of worms in the body. He questioned the need of worms in the body [71]: "Do worms produce disease from an *excess* in their *number*, and an error in their place, in the same manner that blood, bile, and air produce disease from an *error* in their place, or from *excess* in their *quantities*?" Rush had some anecdotes that might prove his ideas that worms were related to disease processes. He had noted that in the cases of small pox and measles, children also suffered from worms [72]. Rush even thought that worms were a good thing in the body of children, as the worms might correct the balance of their diet, and children got sick without the appearance of worms. The link between fever and worms was also not agreed upon by physicians at the time, and Rush postulated that worms are a part of fevers, not the cause of them [73].

The key to health was a humoral maxim of balance, and that balance was accomplished through chemical substances. Anthelmintic chemicals were the tools to ridding children of diseases of the nerves or other long-term diseases, and Rush engaged in chemical research to test worm-killing drugs, which he published in article form for other chemists and physicians. Rush even performed chemical experiments where he subjected earthworms to chemicals in the hopes of finding anthelmintic drugs [74]. He also observed that when anthelmintic drugs are taken into the body, they are often transformed by the stomach, which mutes their efficacy. One of the key problems in humoral theory was how drugs interreacted in the body, which would remain a question until the arrival of sulfa drugs around the early twentieth century [75, 76].

3.5 The American Revolution: Chemistry and Humoral Theory

Benjamin Rush served the American Revolutionary government as both a doctor and as someone who was interested in the politics of a new government. He had a lot of challenges during the Revolution: inconsistent leadership, lack of resources, and problems that he created for himself. Rush memorialized many of his experiences being a doctor in the Revolutionary government. Much like the continual change of military leadership in the early years of the American Revolution, the Medical Department of the Continental Army was also a revolving door of leadership. In 1777, John Morgan, one of Rush's mentors, was fired from directing the Medical Department by Congress. Rush was appointed to the position to replace Morgan and served in 1777 as the Surgeon General in the middle area, as the colonies were divided into three districts. While teaching at the University of Pennsylvania, Rush treated many injured soldiers as many of the battles of the Northern campaigns of the American Revolution took place in the Philadelphia area.[8] Rush seemed to maintain both his teaching and military service simultaneously.

Rush had previously served on the Medical Committee which was in charge of medical care of the Continental Army [77, 78]. Rush had been active in taking care of soldiers during the war, as he had been active in the Battles of Trenton and Princeton. Rush had a lot of problems, from having too many patients to care for, as well as a lack of supplies, and there were several diseases that broke out in the patriot forces [79].

Rush blamed William Shippen, who had become the Director General of all the American patriot hospitals, and worked in a covert campaign to blame Shippen for the lack of supplies and problems in the hospital system. In the process, Rush worked to clear his mentor, John Morgan, of any blame, as Morgan had been removed by Congress for the problems with medical supplies. Morgan and Rush were eventually successful at getting Shippen removed as well. Rush was active in the Medical Department until he resigned after news emerged that Rush was covertly criticizing the leadership of General George Washington [77, 78].

In his time in the Medical Department, Rush let his chemical and humoral ideas influence suggestions for good health that he promoted to soldiers. Rush published directives for soldiers maintaining good health. In 1777, he published that [80], "Fatal experience has taught the people of America the truth of a proposition long since established in Europe, that a greater proportion of men perish with sickness in all armies than fall by the sword." Rush had been concerned for the health of the Continental Soldiers, as he had seen disease break out and Army leadership like George Washington had complained about the problems related to diseases and medical care. Encampments like Valley Forge received a lot of complains from Washington, but Congress was never able to improve conditions for soldiers.

Rush tried to focus on four main areas of improvement for soldiers. These notes of improvement included changes in the dress of soldiers, cautions regarding the diet of soldiers, focuses on cleaning, and improvements the conditions of encampments in general. Most of Rush's recommendations were based on the worry of Miasma. Miasma theory was a concept of contagious disease in humoral theory, where debris would spread and cause illness, or irritate the fluids of the body [80]. Miasmas were often associated with swamps and other areas of putrefying bodily material that caused some unknown changes in the body's humors [81]. The liquids of the body spread miasma according to Rush and linen clothing, which was all that the Continental Congress could provide to its soldiers, was not something that Rush wanted to continue. He admonished that [82], "It is a well-known fact that the perspiration of the body, by attaching itself to linen and afterwards by mixing with rain, is disposed to miasmata, which produce fevers." Rush thought that flannel shirts would spread less disease and increase the health of the soldiers.

[8] A good source of Rush's wartime service can be found at: https://www.battlefields.org/learn/biographies/benjamin-rush.

The diet of the soldiers was also important in a humoral system. Most soldiers ate a lot of vegetables, and Rush encouraged that they be thoroughly cooked as he thought they would cause illness by literally decomposing in soldiers [83, 84]. Fresh food rotting in the soldier's body was a large concern to Rush and he did want them to sparingly eat meat with meals. He also encouraged them to consume drugs preemptively in the event of disease, such as Jesuit's bark, as he also thought meat and fevers were somehow related. He emphasized the importance of Jesuit's bark in keeping the soldier's health safe: "if every tree on the continent of America produced Jesuit's bark, it would not be sufficient to preserve or to restore the health of soldiers who eat two or three pounds of flesh in a day." Rush's temperance against alcohol was also well known [85]. He was against alcohol because it manipulated the spirits too much in seasons and allowed the weather to harm their constitutions. He suggested vinegar as it regulated the fluids of the body, as vinegar protected the body against decomposing foods and [86], "it moreover calms the inordinate action of the solids which is created by hard duty." However, there were some cases of watered-down rum that could be helpful to extremely tired soldiers.

Soldiers, Rush encouraged, needed to wash at minimum two to three times during the week, and he really emphasized following this schedule in the summer months, adding that cold baths were needed in order to maintain health. Soldiers not only had to wash their bodies but their food prior to consumption. Disease was also avoided by soldiers if they did not gather into crowds. They also needed to sleep in the sun and all garbage and refuse needed to be removed regularly from the camp.

Sickness was also spread by the winds, and Rush was a strong follower of humoral theory, and believed that the planning of camps needed to know how winds were spreading fever [87, 88],

> Sometimes it may be necessary to encamp an army upon the side of the river. Previous to this step, it is the duty of the quartermaster to inquire from what quarter the winds come at the season of his encampment. If they pass across the river before they reach his army, they will probably bring with them the seeds of bilious and intermitting fevers, and this will more especially be the case in the fall of the year.

Rush noted that the British followed closely the nature of winds and pursued good airs when panning for a camp. Not paying attention to issues that effected the soldier's humoral balance could be detrimental to any army.

Generals seemed to agree with Rush's recommendations, as General Nathaniel Greene endorsed Rush's ideas. Though Rush and Greene publicly got along, Rush criticized Greene in his personal papers, thinking Greene was a leader without discipline [89]. Rush also had problems both publicly and privately being part of the army medical system. During the Revolution, Rush also treated cases of tetanus with humoral interventions, like mercury and blistering.

After the war, Rush would turn his attention to the epidemic that descended upon Philadelphia. Yellow fever was especially difficult for Philadelphia as many of its citizens died from a disease that physicians felt that they had no answer, in a similar manner of the COVID-19 epidemic that occurred in the United States in 2020. Rush again, used his chemical and humoral framework to attempt to understand the disease and help his patients.

3.6 Yellow Fever, Blood Letting, and Humoral Chemistry

It has been estimated by some historians that the loss of life in the yellow fever epidemic in 1793 killed one out of every ten people in Philadelphia at that time. Yellow fever was much like bubonic plague, where there were several waves of it from the seventeenth and eighteenth century [90, 91]. The skin in yellow fever turned a distinctive yellow, as well as other parts of the body. Rush recorded in his autobiographical writing the sadness the epidemic produced, and it still made him sad several decades later [92]. Yellow fever was both professionally and personally difficult for him, as his ideas about treating yellow fever would have him removed from the College of Physicians that was based in Philadelphia. Rush had been a founder of the college and he found himself alone on the most extreme end of the humoral practice spectrum [93].

Rush theorized that yellow fever developed through miasmas that existed in the world entering the body and causing a response. Miasmas were produced by the literal garbage the piled up on the streets of Philadelphia [93–95]. Rush also thought that miasmas were bread through bad air and water [81]. There were also problems with sewer run off, as well as dirty basements and yards in the city, making for ripe fever conditions. Outhouses also produced additional bad airs that were linked to producing disease. Not all doctors believed that yellow fever came from local sources like Rush, and in fact blamed his aggressive bloodletting program for putting patients at risk, especially when he combined high amounts of medicines that purged the body.

However, Rush believed that external forces acted on the fluids or humors of the body, such as weather and bad airs. He summarized all the external actors on the body that produced yellow fever [96], "... An exciting cause, such as great heat, cold, fatigue from riding, walking, swimming, gunning, or unusual labour, intemperance in eating or drinking, ice creams, indigestible aliment, or violent emotion of the mind."

Though Rush spent a lot of time explaining to readers why people suffered from yellow fever, he also attempted to explain why some people avoided the fever entirely. As a strong adherent to humoral theory he argued that there were simply constitutional differences between people that prevented them from getting sick. Perhaps they knew to run from bad airs. Or abstained from alcohol. The humoral answer to spreading disease was the idea of "pecctant humors" that were toxic or expelled humors that spread disease, or people could be simply trapped in a bedroom with no flowing air [97]. Extreme emotional states could also spread disease.

Rush was optimistic that a solution existed in nature. As a humoral theorist, however, a drug-based solution did not entirely fit into humoral pathology, but the concept was being stretched by Rush and other medical chemists. He wrote of a natural cure [98], "to every natural evil, Heaven has provided an antidote, and it is not more certain, that houses are preserved from the destructive effects of lighting by metal conductors, than our cities might be preserved, under the usual operation of the laws of nature, from the yellow fever by *cleanliness*."

Rush gradually became more paranoid that he was being undermined by colleagues, especially given their disapproval of his treatment of yellow fever. The historian Paul E. Kopperman contextualized Rush's interpersonal difficulties in his article about the history of bloodletting in yellow fever [99],

> He was high strung and combative, easily offended and slow to forgive. During the course of the epidemic, he published several letters attacking physicians who followed methods other than depletion and he showed little collegiality toward them, providing at most a patina of respect. He had some allies, but he felt betrayed by friends who adhered to other practices and he also feared that rival methods would win over the people, thousands of whom would die in consequence.

Rush's paranoia continued to make him charge his detractors with secretly deterring students from studying with him and undermining all his medical ideas. But Rush himself thought that he did a good job treating yellow fever, writing proudly in his autobiography that [100]:

> The success which attended the remedies which it pleased God to make men the instrument of introducing into general practice in the treatment of the fever of 1793 produced a sudden combination of all who had been either publickly or privately my enemies, and the most violent and undisguised exertion to oppose and discredit those remedies.

In some cases, Rush worked to mended relationships with critics and put personal disagreements aside, and he nursed and cared for many of those same physicians who had been critical of him when they fell ill with yellow fever [101]. By looking at Rush's use of bloodletting, I think that historians can view it as a type of chemical intervention, though Rush himself might not have made that association. However, the context of why Rush was making interventions into the body is line with chemical ideas about the body and how to manage the body's fluids.

Rush's magnum opus in regards to bloodletting was *A Defense of Blood Letting* that explained all manner of uses for bloodletting, include treating urinary stones. Bloodletting functioned as a method to remove stimulation from the body. In the case of fever, removing the hot blood cooled the patient down. And any part of the body could become overly excited, and it was the job of the physician to balance the stimulation of the body. Bloodletting was one intervention to balance out over stimulation, which could include temperature, the light, moving the body, food, and emotions. Solutions were essentially the opposite intervention of the problem. For instance, if a patient was suffering from diet, they were supposed to avoid food, or if they were being bothered by too much light they needed to find seclusion in the darkness. If the patient was too amorous, they needed to have fear put in them. There were also medications that were thought to act on the over stimulated power of the body, such as foxglove or salt, among other medications [102]. The doctor could add stimulants to the system in order to balance it as well, such as alcohol or mercury. The physicians also had baths and blistering to further stimulate the body. Medical philosophers like Rush thought that alcohol was capable of stimulating effects, which is different than current physicians' conception of alcohol as a strong depressant.[9]

The desire to defend bloodletting came from the harsh criticisms he received during the yellow fever outbreak of 1793, and Rush wanted to defend his actions, as well as his philosophy of medicine. Bloodletting released the extreme stimulation and inflammation that fever produced in the body, and as it spread to over organ systems. Rush believed that fever was produced in the body by something outside it, and bloodletting was a part of nature. He used the example of female menstruation. He contextualized the natural loss of blood in the body [103], "She frequently pours the stimulating and oppressing mass of blood into the lungs and the brain; and when she finds an outlet for it through the nose, it is discharged either in such a deficient or excessive quantity, as to be useless or hurtful." Bloodletting was resetting the body, and the physician could do it more scientifically than the natural process. And bloodletting was known to cure tetanus and other diseases. It was also crucial for medications to work, as it cleared away for them to spread into the body [104]. He criticized objectors to bloodletting as using the same arguments that were known to Galen in the ancient past.

The extent to apply bloodletting was vast to Rush, as he had bled his own children at early ages, ensuring their proper health in adulty hood. Women, Rush advocated, should also be bleed during menstruation as well so that they avoided fevers related to inflammation [105]. The amount that Rush advocated for was a large amount of blood [105], "Formerly the natural discharge from the uterus was trusted to remove a fever contracted during the time of menstruation. But what relief can the discharge of four or five ounces of blood from the uterus afford, in a fever which requires the loss of 50, or perhaps of 100 oz to cure it?".

Much like chemists advocating methods to remove the blockages of the body. Rush saw bloodletting as a cure to remove concretions and other blockages in the body's fluid systems. In cases of pregnancy, bloodletting would keep the patient from experiencing compilations. Rush promoted the idea that [106]:

> The severity of the pains in these cases created a disease, which prevented internal congestion in the womb. Bleeding, by depleting the uterus, obviates at once both disease and congestion. Its efficacy is much aided by means of glysters, which, by emptying the lower bowels, lessen the pressure upon the uterus.

[9] See Hirsh J (1949) Enlightened Eighteenth Century Views of the Alcohol Problem. J Hist of Med Allied Sci 2:230–236, Katcher BS (1993) Benjamin Rush's Educational Campaign Against Hard Drinking. Am J Pub Health 83:273–281, and Levine HG (1984) The Alcohol Problem in American: From Temperance to Alcoholism. Br J Addict 79:109–119.

As a chemist, Rush was very precise about the amount of blood that was recommended to remove from the patient. Rush knew that most people had anywhere around 25–28 lbs of blood in their body. The precision from chemistry suggested to Rush that a patient only needed to lose four to five lbs of blood from their body in order to restore the system and improve blood circulation [107]. He made an analogy with wine as stimulating the circulating system [108]:

> Thus very small portions of light, and sound, are sufficient to excite vision and hearing in an inflamed, and highly excitable state of the eyes and ears. Thus too, a single glass of wine will often produce delirium in a fever in a man, who, when in health, is in the habit of drinking a bottle every day without having his pulse quickened by it.

Rush shammed other physicians who did not have the precision of removing blood, such as in the case of a doctor simply removing 320 oz of blood over the long period of almost a month, or another case of a physician trying to treat a woman for hysteria that he only took 1020 oz over the period of almost twenty years [109]. The criticism was based more in the fact that those doctors not enough blood over too long a period of time, as Rush did not believe in problems related to taking too much blood too quickly, as he thought blood naturally regenerated quickly [110]. Rush advised no caution when breathing the vein, even when the patients lacked color in their faces [111]. And he thought that bleeding could be used to help patients transition out of life and ease their suffering [112]: "It belongs to this remedy, in such cases, to ease pain, to prevent convulsions, and compose the mind, to protract the use of reason, to induce, sleep, and thus to smooth the passage out of life."

3.7 Rush and His Own Final Bloodletting: Conclusion

In 1813, Benjamin Rush would succumb to fever, a bookend to all of his research and writing about fevers. He would also choose bloodletting as a preferred method to treat the fever. Rush took tea after making the round with his patients on April 14. He was seized with a chill, and he was bled. The bleeding of 10 oz of blood did not cure Rush, and he called for two doctors. The first doctor was Dorsey, and then Dr. Phillip Physic, another famous physician of Philadelphia history was called in to examine Rush [113]. Both Dorsey and Physic disagreed about what was making Rush sick. Dorsey thought that Rush might be suffering from typhoid fever, while Physic disagreed, and argued that Rush was suffering from tuberculosis [8]. Physic was permitted by Rush to remove more blood, this time through the process of cupping [114]. Though Rush initially felt better, he died on April 19. News of Rush's death ran in many publications and was acknowledge by many leaders in turn of the nineteenth century America [115].

In historicizing Rush's life and ideas about medicine, humoral medicine both defined his life and his practice. Understand the body's fluids and how they contributed to disease was the key to medicine, as far as Rush was concerned.

Understanding the fluids of the body revealed how irritation and outside miasmas caused disease processes in the body, and Rush hoped that by adjusting those fluids he could improve the course a disease was taking in the body. Chemistry, as Rush was a trained chemist, could contribute to further understanding how the body's humors cause disease and how nature (especially miasmas) manipulated those humors. Rush had chemistry at the heart of his medical practice, as well as humoral pathology. Better chemical analysis could provide better guidance for bloodletting and medicines that removed worms from the body. Chemistry was used not only in medicine, but in Rush's theories about race and the need for the practice of slavery.

Rush's medical authority regarding the fluids of the body, especially in cases of blockages, such as urinary stones. In 1793, Benjamin Rush received a letter from John Willday, likely a merchant from Philadelphia [115, 116]. Rush published the letter from Willday in the *Transactions of the College of Physicians of Philadelphia* and entitled it "Account of the Successful Application of Cold Water to the Lumbar Region in Calculous Cases." In 1788, Willday wrote that he was having severe pains that he attributed to the "gravel," which referred to urinary stones or related stones in the eighteenth century [117]. He suffered from bodily stones quarterly throughout the year. He had severe pain as the stones often broke up into shards, which he felt threated his life at some point. But thanks to [118], "the blessing of God, *I have* discovered a method of preventing those terrible fits of the stone, so that I have not had one of them for near two years, and my remedy has been tried by another person with equal success." In order to find this method out, Willday had to read extensively about the problem of the stone from every writer he could gather. He found that cold baths, as he noted that most author's encouraged relaxation, and after passing the last bit of stone from his body, he attempted a cold bath to determine if it could help. He described the technique [119],

> I placed a bason of cold water on a night chair, and on getting out of bed every morning stripped myself naked, and placing myself over it, I applied a large wet spunge to the small of m back two or three times, and I sometimes applied it to my fundament and the genitals; the consequence of which has been, that the vessels of my kidneys have been contracted and strengthened, so that they have expelled the sand and gravel as fast as it formed or generated, so that though I pass as much sand and gravel as before, I have never had a bad fi since I began with the above method which is now nearly two years.

Willday sent the letter in the hopes that it would provide aid to someone "in America," and knew to send it to Rush, as Rush was a major disseminator of medical knowledge. Publishing it in a major medical and scientific journal like the *Transactions of the College of Philadelphia* would have the letter read by many readers [120]. Rush however, would later have a falling out with the College during the 1793 Yellow Fever outbreak mentioned in this chapter. Rush's interest in understanding the fluids of the body and the problems related to those blockages could be the reason why Willday sent Rush the letter. The letter also represents the anecdotal or hint culture of medicinal chemistry in its focus on understanding the body's fluids and humoral pathology [117].

Edward Darrell Smith, the subject of the next chapter, continued to publish medical and chemical anecdotes meant to help suffers from bodily stones and those who had bodily fluid problems. Rush's work was based in major metropolitan areas like Philadelphia and his training in Edinburgh. However, the chemical and humoral pathological ideas of Rush filtered through his courses in Philadelphia and were transmitted through journal literature that reached practitioners that would travel to the rural areas of the American South and continue to research similar ideas to Rush, but also continue the chemical research of looking into how the blockage of the body's fluids were tied to disease processes, like urinary stones. Edward Darrell Smith, a physician trained in Philadelphia while Rush was a professor at the University of Pennsylvania. The next chapter will examine Smith's life and his chemical investigations into urinary stones.

References

1. Rush B (1805) Medical inquiries and observations, vol 2, 2nd edn. T & G Palmer, Philadelphia, p 259
2. Thomas DP (2014) The demise of bloodletting. J R Coll Phys Edinb 44:72–77
3. Rush B (1805) Medical inquiries and observations, vol 2, 2nd edn. T & G Palmer, Philadelphia, p 260
4. Fried S (2018) Rush: revolution, madness, and Benjamin Rush, the visionary doctor who became a founding father. Crown, New York
5. Powers J (2012) Inventing chemistry: Herman Boerhaave and the reform of the chemical arts. University of Chicago Press, Chicago
6. Powers J (2012) Inventing chemistry: Herman Boerhaave and the reform of the chemical arts. University of Chicago Press, Chicago, p 109
7. Christie JR (1994) The historiography of chemistry in the eighteenth century: Herman Boerhaave and William Cullen. Ambix 41:4–19
8. Miles W (1953) Benjamin Rush, Chemist. Chymia 4:37–44
9. Good HG (1918) Benjamin Rush and his services to American education. Witness Press, Berne
10. Binger C. Revolutionary doctor: Benjamin Rush 1746–1819. W. W. Norton & Company, New York
11. North RL (2000) Benjamin Rush, MD: assassin or beloved healer? Proc Bayl Univ Med Cent 13:45–49
12. Klickstein HS (1953) A short history of the professorship of chemistry of the university of Pennsylvania school of medicine: 1765–1847: with an account of the controversy over the chair, 1809–1818. Bull Hist Med 27:43–68
13. Stewart J (2012) The reality of phlogiston in Great Britain. Hyle 18:175–194
14. Stewart J (2020) Chemistry and slavery in the Scottish enlightenment. Ann Sci 77:155–168
15. Donovan J (1975) Philosophical chemistry in the Scottish enlightenment: the doctrines and discoveries of William Cullen and Joseph Black. Edinburgh University Press, Edinburgh
16. Sacin MB, Abrahams HJ (1956) Benjamin Rush's course in chemistry at the young ladies' academy. J Frank Inst 262:425–435
17. Lambert PF (1972) Benjamin Rush and American independence. Penn Hist 39:443–454
18. Jerber LK (1980) Women and the republic: intellectual and ideology in revolutionary America. University of North Carolina Press, Chapel Hill
19. Rush B (1811) Sixteen introductory lectures of courses upon the institutes and practice of medicine, with a syllabus of the Latter. Bradford and Inskeep, Philadelphia

20. Rush B (1811) Sixteen introductory lectures of courses upon the institutes and practice of medicine, with a syllabus of the latter. Bradford and Inskeep, Philadelphia, pp 223–348
21. Rush B (1811) Sixteen introductory lectures of courses upon the institutes and practice of medicine, with a syllabus of the latter. Bradford and Inskeep, Philadelphia, p 348
22. Sussman RW (2014) The myth of race: the troubling persistence of an unscientific idea. Harvard University Press, Cambridge
23. Mallet JW (1876) Notes of work by students of practical chemistry in the laboratory of the university of Virginia No. V. Chem News Oct 27:147–169, 190–191, see p 180
24. D'Elia DJ (1969) Dr. Benjamin Rush and the Negro. J Hist Idea 30:413
25. D'Elia DJ (1974) Benjamin Rush: philosopher of the American revolution. Trans Am Phil Soc 64(part 5)
26. McCandless P (2014) Slavery, disease, and suffering in the southern lowcountry. Cambridge University Press, New York
27. Wheeler R (2000) The complexion of race: categories of difference in eighteenth-century British culture. University of Pennsylvania Press, Philadelphia
28. Curran SA (2011) The anatomy of blackness: science and slavery in the age of enlightenment. Johns Hopkins University Press, Baltimore
29. Malcolmson C (2014) Studies of skin color in the early royal society: Boyle, Cavendish. Routledge Press, London, Swift
30. Henze B (2004) Scientific definition in rhetorical formations: race as 'permanent variety' in James Cowles Prichard's ethnology. Rhetoric Rev 23:311–331
31. Chiles KL (2014) Transformable race: surprising metamorphoses in the literature of early America. Oxford University Press, pp 17–18
32. Dain B (2002) A hideous monster of the mind: American race theory in the early republic. Harvard University Press, Cambridge, pp 41–42
33. Smith SS (1810) An essay on the causes of the variety of complexion and figure in human species. J. Simpson and Co, Williams and Whiting, and L. Deare, New Brunswick
34. Rush (1799) Observations to favour a supposition that the black color (as it is called) of the Negroes is derived from the leprosy. Trans Am Phil Assoc 4:289–297
35. Plummer BL (1970) Benjamin Rush and the Negro. Am J Psych 127:93–98
36. Herschthal E (2017) Antislavery science in the early republic: the case of Dr. Benjamin Rush. Early Amer Stud 15:274–307
37. Rush (1799) Observations to favour a supposition that the black color (as it is called) of the Negroes is derived from the leprosy. Trans Amer Phil Assoc 4:289
38. Motherby G (1776) A new medical dictionary: or, general repository of physic. J. Johnson, London
39. Rush (1799) Observations to favour a supposition that the black color (as it is called) of the Negroes is derived from the leprosy. Trans Am Phil Assoc 4:290
40. Ward DC (2004) Charles Wilson Peale: art and selfhood in the early republic. University of California Press, Berkley
41. Villarosa L (2019) Myths about physical racial differences were used to justify slavery—and are still believed by doctors today. NY Times
42. Rush (1799) Observations to favour a supposition that the black color (as it is called) of the Negroes is derived from the leprosy. Trans Am Phil Assoc 4:292
43. Rush (1799) Observations to favour a supposition that the black color (as it is called) of the Negroes is derived from the leprosy. Trans Am Phil Assoc 4:294
44. Delacy M (2017) Contagionism catches on: medical ideology in Britain, 1730–1800. Springer, New York
45. Rush (1799) Observations to favour a supposition that the black color (as it is called) of the Negroes is derived from the leprosy. Trans Am Phil Assoc 4:295
46. Curan A (2009) Rethinking race history: the role of Albino in the French enlightenment life sciences. Hist Theory 48:151–179
47. Rush (1799) Observations to favour a supposition that the black color (as it is called) of the Negroes is derived from the leprosy. Trans Am Phil Assoc 4:296

48. Brooks WJ (1803) History of the progress and present state of animal chemistry. J. Johnson, London
49. Brooks WJ (1803) History of the progress and present state of animal chemistry, vol 2. J. Johnson, London, p 224
50. Hannaford I (1996) Race: the history of an idea in the west. The Woodrow Wilson Center Press, Washington
51. Brooks WJ (1803) History of the progress and present state of animal chemistry, vol 2. J. Johnson, London, p 225
52. Brooks WJ (1803) History of the progress and present state of animal chemistry, vol 2. J. Johnson, London, pp 227–228
53. Stephan N (1982) Idea of race in science: Great Britain, 1800–1960. Springer, New York
54. Levere T, Stewart L, Torrens H, Wachelder J (2017) The enlightenment of Thomas Beddoes: science, medicine, and reform. Routledge, New York
55. Beddoes T (1794) Considerations on the medicinal use of factitious airs, and on the manner of obtaining them in large quantities in two parts. Bulgin and Rosser, Bristol
56. Beddoes T (1794) Considerations on the medicinal use of factitious airs, and on the manner of obtaining them in large quantities in two parts. Bulgin and Rosser, Bristol, pp 32–33
57. Jefferson T, Peden W (eds) (1955) Notes on the state of Virginia. University of North Carolina, Chapel Hill, p 138
58. Jefferson T, Peden W (eds) (1955) Notes on the state of Virginia. University of North Carolina, Chapel Hill, p 139
59. Rush B (1769) An account of the effects of the Stramonium, or thorn-apple. Trans Am Phil Soc 1:318–322
60. Cox FEG (2002) History of human parasitology. Clin Microbiol Rev 15:595–612
61. Chakrabarti P (2010) Materials and medicine: trade, conquest and therapeutics in the eighteenth century. Manchester University Press, Manchester
62. Rush B (1769) An account of the effects of the stramonium, or thorn-apple. Trans Am Phil Soc 1:319–320
63. Rush B (1769) An account of the effects of the stramonium, or thorn-apple. Trans Am Phil Soc 1:319
64. Balzer W, Eleftheriadis A (1991) A reconstruction of the hippocratic humoral theory of health. J Gen Phil Sci 22:207–227
65. Sullivan RB (1994) Sanguine practices: a historical and historiographic reconsideration of heroic therapy in the age of rush. Bull Hist Med 68:211–234
66. Rush B (1769) An account of the effects of the stramonium, or thorn-apple. Trans Am Phil Soc 1:322
67. Rush B, Butterfield LH (eds) (1951) To Thomas Jefferson: an account of the sugar maple tree. Princeton, University Press, Princeton, pp 587–599
68. Rush B, Butterfield LH (eds) (1951) To Thomas Jefferson: an account of the sugar maple tree. Princeton, University Press, Princeton, p 594
69. Rush B (1805) Observations upon worms in the alimentary canal, in medical inquiries and observations 1. Hopkins and Earle, Mathew Carey, Johnson and Warner, Kimber and Conrad, Bradford and Inskeep, Thomas and William Bradford, Benjamin and Thomas Kite, and Bennett and Walton, Philadelphia, pp 215–235
70. Rush B (1805) Observations upon worms in the alimentary canal, in medical inquiries and observations 1. J. Conrad & Co., Philadelphia, p 217
71. Rush B (1805) Observations upon worms in the alimentary canal, in medical inquiries and observations 1. J. Conrad & Co., Philadelphia, p 218
72. Rush B (1805) Observations upon worms in the alimentary canal, in medical inquiries and observations 1. J. Conrad & Co., Philadelphia, p 219
73. Rush B (1805) Observations upon worms in the alimentary canal, in medical inquiries and observations 1. J. Conrad & Co., Philadelphia, p 220
74. Rush B (1805) Observations upon worms in the alimentary canal, in medical inquiries and observations 1. J. Conrad & Co., Philadelphia, pp 221–224

75. Holland BK (ed) (1996) Prospecting for drugs in ancient and medieval European texts: a scientific approach. Harwood Academic Publishers, Amsterdam
76. Foster GM (1994) Hippocrates' Latin American legacy: humoral medicine in the new world. Gordon and Breach Scientific Publishers, Amsterdam
77. Gillet MC (1990) The army medical department: 1775–1818. Center of Military History United States Army, Washington
78. Reiss O (1998) Medicine and the American revolution: how diseases and their treatments affected the colonial army. McFarland & Company, Jefferson
79. Fenn EA (2002) Pox Americana: the great smallpox epidemic of 1775–82. Hill and Wang, New York
80. Johnson S (2006) The ghost map: the story of London's most terrifying epidemic and how it changed science, cities, and the modern world. Riverhead, New York
81. Driggers E (2022) "… [T]he movement of a celestial system than a human invention:" Abram Blanding and bringing water to Columbia. Water Hist 14:21–40
82. Rush B, Washington G (1951) 1777 in Butterfield L. H. ed. (1951) Letters of Benjamin Rush: Volume I: 1761–1792. Princeton University Press, Princeton, pp 180–182
83. Rush B (1951) To the Officers in the Army of the United American States: Directions for Preserving the Health of Soldiers in Butterfield L. H. ed. (1951) Letters of Benjamin Rush: Volume I: 1761–1792. Princeton University Press, Princeton, pp 140–147
84. Mathias M, Moore AM (eds) (2018) Gut feeling and digestive health in nineteenth century literature, history, and culture. Springer, New York
85. Katcher BS (1993) Benjamin Rush's educational campaign against hard drinking. Am J Pub Health 83:273–281
86. Rush B (1951) To the Officers in the Army of the United American States: Directions for Preserving the Health of Soldiers in Butterfield L. H. ed. (1951) Letters of Benjamin Rush: Volume I: 1761–1792. Princeton University Press, Princeton, p. 142.
87. Rush B To the Officers in the Army of the United American States: Directions for Preserving the Health of Soldiers in Butterfield L. H. ed. (1951) Letters of Benjamin Rush: Volume I: 1761–1792. Princeton University Press, Princeton, p. 144.
88. Reiss O (1998) Medicine and the American revolution: how diseases and their treatments affected the colonial army. Jefferson NC, McFarland
89. Greene N to B. Rush, 3rd of May 1777 in Showman RK ed. (1980) The Papers of Nathaniel Greene. University of North Carolina Press, Chapel Hill, p. 68
90. Duffy J (1953) Epidemics in colonial America. Louisiana State University Press, Baton Rouge
91. Crosby MC (2006) The American plague: the untold story of yellow fever, the epidemic that shaped your history. Berkley Books, New York
92. Rush B (1948) Autobiography of Benjamin Rush. In: Corner GW (ed). Princeton University Press, Princeton, pp 95–96
93. Middleton WS (1928) The yellow fever epidemic of 1793 Philadelphia. Ann Med Hist 10:434–450
94. Powell JH (1949) Bring out your dead; the great plague of yellow fever in Philadelphia in 1793. University of Pennsylvania Press, Philadelphia
95. Kopperman PE (2004) "Venerate the lancet:" Benjamin Rush's yellow fever therapy in context. Bull Hist Med 78:539–574
96. Rush B (1799) Observations upon the origin of the malignant bilious, of yellow fever in Philadelphia, p 6
97. Child III PW (2021) The dean disordered: Jonathan Swift and the humoral body. Thesis for M A Johns Hopkins, Baltimore
98. Rush B (1799) Observations upon the origin of the malignant bilious, of yellow fever in Philadelphia, p 26
99. Kopperman PE (2004) "Venerate the lancet:" Benjamin Rush's yellow fever therapy in context. Bull Hist Med 78:567

100. Rush B (1948) Autobiography of Benjamin Rush. In: Corner GW (ed). Princeton University Press, Princeton, p 97
101. Hedges WL (1973) Benjamin Rush, Charles Brockden Brown, and the American plague year. Early Am Lit 7:295–311
102. Rush B (1796) A defense of blood-letting as a remedy for certain diseases. In: Medical inquiries and observations, vol IV. Thomas Dobson, Philadelphia, pp 183–254
103. Rush B (1796) A defense of blood-letting as a remedy for certain diseases. In: Medical inquiries and observations, vol IV. Thomas Dobson, Philadelphia, p 188
104. Rush B (1796) A defense of blood-letting as a remedy for certain diseases. In: Medical inquiries and observations, vol IV. Thomas Dobson, Philadelphia, p 194
105. Rush B (1796) A defense of blood-letting as a remedy for certain diseases. In: Medical inquiries and observations, vol IV. Thomas Dobson, Philadelphia, p 199
106. Rush B (1796) Medical inquiries and observations, vol III. Thomas Dobson, Philadelphia, p 355
107. Rush B (1796) A defense of blood-letting as a remedy for certain diseases. In: Medical inquiries and observations, vol IV. Thomas Dobson, Philadelphia, p 234
108. Rush B (1796) A defense of blood-letting as a remedy for certain diseases. In: Medical inquiries and observations, vol IV. Thomas Dobson, Philadelphia, pp 234–235
109. Rush B (1796) A defense of blood-letting as a remedy for certain diseases. In: Medical inquiries and observations, vol IV. Thomas Dobson, Philadelphia, p 237
110. Waring JI (1961) The influence of Benjamin Rush on the practice of bleeding in South Carolina. Bull Hist Med 35:230–237
111. Rush B (1796) A defense of blood-letting as a remedy for certain diseases. In: Medical inquiries and observations, vol IV. Thomas Dobson, Philadelphia, p 238
112. Rush B (1796) A defense of blood-letting as a remedy for certain diseases. In: Medical inquiries and observations, vol IV. Thomas Dobson, Philadelphia, p 241
113. Bringer Revolutionary Doctor, pp 294–296
114. North RL (2000) Benjamin Rush, M. D.: assassin or beloved healer? Proc Bayl Univ Med Cent 13:46
115. Willday J (1793) Account of the successful application of cold water to the lumbar region in calculous cases. In a letter to Benjamin Rush, M. D. Trans Coll Phys Phil 1:76–78
116. Nelson W (1903) Extracts from American newspapers: relating to New Jersey: 170-1775, vol 25. The Call Printing and Publishing Co., Paterson, p 297
117. Driggers EA (2020) "A clue in the labyrinth:" the medico-chemistry of the stone in the life of Edward Darrell Smith. J Med Bio 28:232–245
118. Willday J (1793) Account of the Successful Application of Cold Water to the Lumbar Region in Calculous Cases. In a letter to Benjamin Rush, M. D. Trans Coll Phys Phil 1:76
119. Willday J (1793) Account of the successful application of cold water to the lumbar region in calculous cases. In a letter to Benjamin Rush, M. D. Tran Coll Phys Phil 1:77–78
120. Bell JW (1988) The college of physicians of Philadelphia: a bicentennial history. Since History Publications, Canton

Medico-Chemistry and the American South: The Life of Edward Darrell Smith

4

Edward Darrell Smith (1775/1778–1819) was a chemist, physician, and both a rural and urban American during the early nineteenth century. He was a physician who was sought out for assistance in his Western South Carolina community. Smith had been born in the elite city of Charleston in the eighteenth century, seen a father pass away from disease on a prison boat during the American Revolution, learned medicine in his home state of South Carolina, then traveled to Philadelphia to study at the University of Pennsylvania, before returning home to practice in both urban and rural parts of the state. Smith's career was incredibly interesting, as he shifted his focus from medicine to chemistry in order to improve the problems he saw in medicine. He wanted a better knowledge of chemistry in order to prevent the terrible cases of urinary stones that he treated throughout his career. However, the more chemistry he learned, the more he was convinced that humoral pathology was one of the best medical theories available. Smith's chemical knowledge continued to grow, and he continued to make statements about the high value of humoral theory to medicine. Since his dissertation, Smith had been combining humoral pathology and chemistry to explain how the body and medications worked.

Smith not only took his mix of chemistry and humoral pathology into medical practice, but also into his teaching at the newly established South Carolina College. He also communicated his ideas about medicine and chemistry through the latest scientific and medical journals of the time. Smith was at times a member of a rural community, but his life and engagement demonstrate the extensive nature of communities that were interested in chemistry, humoral pathology, and improving knowledge about the stones of the body.

E. A. Driggers, *Early Nineteenth Century Chemistry and the Analysis of Urinary Stones*, Perspectives on the History of Chemistry,
https://doi.org/10.1007/978-3-031-34973-7_4

4.1 Urological Cases and a Spur to Action

Smith had moved to Western South Carolina, around where the city of current city of Pendleton exists, which is near Clemson University or the modern Greenville-Spartanburg metropolitan area. By 1811, Smith had quit his medical practice and stopped working at his proprietary slave hospital and moved out into rural South Carolina in order to run his own plantation. While Smith was planting in Pendleton, one of his neighbors had a medical emergency and he had occasionally practiced medicine, upon request, in Pendleton [1]. An unnamed man felt his situation with his child was so dire that he traveled from Virginia in November of 1811 to consult with Smith. Smith would publish an account of the medical cases almost nine years later, as he thought that it might help other practitioners faced with a similar case. Smith described the child and their urological problem [1], "… the child was about five years old, that soon after his birth he was observed at times to pass his urine with difficulty, and that this affection continued to increase until about three weeks ago, when a total stoppage took place." The child was suffering greatly, and Smith knew that he needed to act. Smith had been trained in the heroic medical model under the direction of Benjamin Rush [2]. With his prior experiences as guide, he thought that the child might be suffering from a stone [3], "I concluded that a calculus had been expelled from the bladder into the urethra, in which it was so fast wedged, that not a drop of water could pass, and that the fistulous orifice had been formed behind it."

Smith was inspired by cases of urinary stones to pursue chemistry in order to prevent children or other people from suffering from painful cases of stones. Any theory or chemical idea was appealing to Smith in order to prevent stones from forming and causing the patient pain, such as in the case he observed in Pendleton. Smith also wanted to transmit descriptions of cases of the stone that he treated in order to hopefully improve medical practice. He thought that anecdotal information might help another physician more effectively treat a case of the stone. Smith was writing at the turn of the nineteenth century when medical journal literature was expanding and communities formed around the transmission of medical and scientific information [4, 5]. By treating cases of urinary stones over the course of his career, as well as his chemical research, Smith gradually become a supporter of humoral pathology and called for a general return of medical practice to the principles of humoral theory.

This chapter will examine Edward Darrell Smith's life in order to understand how he linked humoral and chemical ideas to the medical treatment of urinary stones. Starting in his medical dissertation, Smith was interested in the chemistry that supported humoral theory to explain medical problems in the body. Throughout his career, he also encountered many cases of urological stones and was inspired to apply new chemical ideas that emerged out of the nineteenth century to treating, or even preventing, urinary stones. The ideas that Smith learned as a medical student also reflected the chemical-based medical ideas of the nineteenth century that were infused with humoral pathology. Smith fits the model of the humoral-based medicine of the nineteenth century, according to the historian

Charles Rosenberg. Rosenberg commented that physicians wanted to control the fluids of the body in order to control disease [6], "The physician's most potent weapon was his ability to 'regulate the sections'—to extract blood, or promote the perspiration, or the urination, or defecation which attested to his having helped the body to regain its customary equilibrium." Rosenberg historicized most diseases as a lack of balance inside the body, and the doctor's job was to return the body back to its normal, balanced state, using either drugs or humoral intervention [7]. There was a therapeutic revolution which prioritized drugs over interventions, but most physicians continued to use this combination of remedies [8, 9].

4.2 Education from North to South: The Life of Edward Darrell Smith

Edward Darrell Smith was trained in this strange mix of humoral medicine, therapeutic drugs, and new ideas from European chemistry [10]. In either 1775 or 1778, depending on the historian or the calendar system, Edward Darrell Smith was born to Mary Elizabeth Stevens and Josiah Smith II (also known as Josiah Smith Jr.) in Charleston, South Carolina [11–15]. Smith traveled from Charleston to the College of New Jersey, which is now currently Princeton University. Smith graduated at the top of his class as valedictorian and stayed at the College of New Jersey to teach and pursue a master's degree. After his education in the North, Smith returned to South Carolina to apprentice in medicine in Charleston. Smith apprenticed under the physician-historian David Ramsay, as well as under William Stevens Smith [16]. He then left Charleston again for Philadelphia, where he studied medicine at the University of Pennsylvania under Benjamin Rush and Benjamin Smith Barton [11]. After graduating from Pennsylvania and receiving his doctorate in medicine, he returned once again to Charleston. In Charleston, Smith tried to profit off of the slave trade and opened a hospital for slaves with another Pennsylvania graduate, John Parker Gough [11].

After Smith graduated from medical school, he started publishing reports regarding urological events. In 1801, he published a case of a man he had treated that suffered from urinary stones [17]. By 1807, Smith had decided to leave the profession of medicine and become a planter in Pendleton, South Carolina, until 1811. Smith observed many cases of stones while still occasionally practicing medicine, such as the case that opened up this chapter. South Carolina College was established in 1801; during the time period, the state capital was moved from Charleston to the new city of Columbia [18, 19]. Smith took up the chair of chemistry there in 1812 [20]. Smith tried to bring the latest chemical research in the classroom, teaching his students everything from electricity to air pumps [21–23]. The records of the Trustees of the South Carolina College have entries regarding the many chemical instruments that Smith purchased for the College [21–23]. However, Smith complained about their reluctance to fully fund the chemical department and complained to famous chemists like Benjamin Silliman [21–23].

In 1819, Smith left Columbia, South Carolina, and traveled West to Missouri. There are many reasons floated as to why Smith left South Carolina College during the summer break. There was a western land company that Smith had invested in, and he was scouting land locations [24]. Smith had written a new will prior to his journey in which he claimed he was journeying west to make more money [11–15]. Smith might have been planning to leave the state as he had become more frustrated with the system of slavery. Smith writes to Benjamin Silliman complaining about the nature of slavery in South Carolina and ultimately moves West in order to improve his health, as Smith himself suffered from urinary stones. He wrote of slavery in South Carolina [25]:

> I have even contemplated a final residence in the country, in the course of two or three years, on the grounds, that my health is injured by a sedentary life, our institution not being conducted in a manner that I can relish, and an ardent desire to escape from a State, so debated by Slavery, as ours is.

However, Smith took ill from a fever at Mr. David Coulter's house in the Missouri territory in 1819 [26]. Most of what is left about Smith's life falls into the category of microhistory. It is by the reconstruction of his ideas through his published articles can historians see the interesting practice ideas of his combining humoral pathology and chemistry to treat the stones of the body [27].

4.3 A Humoral Education: The Dissertation of Edward Darrell Smith

Part of Smith's medical education was the writing of a dissertation. Medical schools at the turn of the nineteenth century required students to write a dissertation on a medical question of their choosing. Previously, European medical schools had their students compose these in Latin. But in America, most dissertations were handwritten in English and sometimes were published for public and professional consumption [28]. Smith ultimately published his dissertation entitled "Being An Attempt to Prove that Certain Substances are Conveyed, Unchanged, Into the Circulation," which was a dissertation that focused on how humoral theory worked in the body. Smith was investigating how drugs and substances are changed and interact within the body's fluid systems [29]. He investigated both chemical- and physiological-based questions in the dissertation. He was bold, as a graduate student, in promoting the value of humoral theory. He argued that humoral theory was useful as a medical theory and still relevant [30], "During the reign of Humoral Pathology, the opinion, that substances were conveyed unchanged into the circulation, was necessarily adopted by the supporters of that doctrine."

Smith contextualized the state of medical theory for the reader that most of the disease of the body originated in the fluids of the body. Drugs worked in the body by either regulating, correcting, or balancing those fluids [30]. But he noted that humoral theory had fallen out of favor with doctors because there was a lack of a good, comprehensive, explanation of how drugs worked in the body. There were two questions or issues that humoral theories struggled with according to Smith. The first question regarded the notion that medications could not be found in the stomach or in the fluids of the body. The second question regarded the transformation of substances into blood. Smith discussed how milk entered into the body and changed. Milk, if it did not change in the body often resulted in death, as he had cited experimental work where milk was inserted raw into the bodies of animals, and they died. The milk was somehow transformed inside the body and turned into the blood. He did not believe that humoral theory should be rejected from the standpoint of sympathy, which was a medical theory that one part of the body affected each other, such as emotional trouble in the mind affecting the stomach [31]. Smith thought that humoral theory was too quickly dismissed.

In order to persuade the reader that substances can be found in the body in their active states, he examines the fluids of the body, which include milk, saliva, urine, as well as chyle, blood, and sweat. He also examines the bones of the body and argues that substances are simply absorbed by various parts and fluids of the body [32]. The absorption of substances into the body explains how medicines work, as they reach the needed parts of the body, and dangerous substances taken into the body also harm the body. Sometimes breast milk even killed, as Smith included a case where a child died from what his nursemaid was consuming in her own diet. The child's death confirmed that substances often remained in the fluids of the body, such as milk [33],

> In attention to diet in a nurse, is often the unsuspected cause of distressing complains in the suckling child. Active medicines, taken into the circulating fluids of a nurse, will affect the child in an alarming manner. Instances of this kind are not are. If by the collection of facts on this subject, any hints may be given, which may lead to the discovery of a solvent of urinary or biliary calculi, it would be of essential service to mankind.

Smith was in conversation with medical authorities from Percival to Boerhaave, as he was a well-read medical student. He cited the work of one of his teachers, Benjamin Smith Barton, who argued that applying turpentine to the skin allows the medicine to transport itself to the bladder and other parts of the body to treat conditions like diabetes. Barton pointed out that the substance did not change, as evidenced by the smell of turpentine in the urine of people that had been working with the substance to clean their hands, like anatomical demonstrators [34]. Good evidence to Smith was how substances changed the fluids of the body. Smith cited that a drug made from longwood turns urine red, while other substances caused the body's fluids to change their taste.

He was especially interested in solutions for treating urinary stones, or "lithotriptic"-related goods and drugs, such as garlic. Garlic, Smith argued, traveled through the body and destroyed stones [35]. Smith also cited the work of Joseph Priestly who had been examining other stone-destroying substances, such as different types of airs and soda [36, 37]. Smith agreed that treating urinary stones with soda was very effective, noting that soda was [34] "... equally efficacious in alleviating the distressing symptoms of nephritis and causing large quantities of gravel to pass of by urine." However, in regards to other lithotriptic remedies, it was unclear if substances like fixed air transformed in the body [38–40]. Other philosophers like Mathew Dobson also advocated for the consumption of soda water to treat the symptoms of the urinary stone [41]. In Smith's opinion, soda water must not change in the body as it treated stones in the urine [42].

Smith did anticipate counterarguments to his ideas and cited the work of Erasmus Darwin as such an objector to his ideas. He summarized Darwin's position that he would object to the idea that substances remained the same in the body, citing the case of diabetes. Darwin believed that substances simply travel directly into the body's bladder and do not transform, noting that because of the high amount of urine produced in the bladder of diabetics, this exceeded the amount that could reasonable be produced in the kidneys. The same was thought about mineral waters, they simply bypassed most of the body, and there must have been some unknown passage where the mineral waters skipped most of the other organs of the body and came to exist in the ureters. However, as a graduate student, Smith could not comment further on the disagreement and wanted other scholars to do more research to solve this problem.

4.4 Medical Cases that Encouraged Smith's Turn to Chemistry

Smith was motivated to engage in chemical research after witnessing some horrendous cases of urinary stones. Early in Smith's career, he treated a man for pain and spasms in his penis and the patient could not pass urine. He is only known to us as "Captain W" who was treated by Smith in Charleston, South Carolina [17]. The patient was healthy, but had problems like fever since March. The patient had "paroxysms" and attempted to treat himself, but Smith does not describe what those treatments entailed [17]. He was not able to urinate until he had something fly out of his urethra, which produced damage to his scrotal area and bloody urine. Smith described the scrotum as "tumify[ing]" with the urine slowing to a standstill [17].

While practicing medicine in Charleston, Smith first encountered the Captain after he visited his uncle, William Smith Stevens [29]. Stevens saw the patient, who was suffering from fever and could not urinate, but there was not any tumification yet. Stevens course of treatment was to give the patients diuretics, in combination with mercury, to produce urine, but these remedies were not helpful, as the Captain had a racing pulse and the continued swelling in his urinary organ increased drastically, described as the [17] "size of a man's head." Stevens consulted with Smith, after Stevens gave the patients a further round of purgatives without success. Smith wrote of the problems the Captain was experiencing [43]:

> Upon Examining the parts, we found that in the most depending operation of the scrotum, on the left side of the rape, a generous spot, of the size of a dollar, had made its appearance, although not visible in the morning. In this spot a lancet was introduced to a considerable depth, and the puncture enlarged afterwards, without the patients being at a sensible of the wounds.

Next, Dr. David Ramsay was called in to consult on the case. At that time, the Captain had been experiencing a lot of fluid buildup at the site of the lancet. The doctors then tried to treat the swelling and the pain with success this time, as they used Jesuit Bark, and other things like "solutions of scacch. saturini" [17]. The Captain reported feeling better and the swelling stopped. Smith theorized at the end of the article as to why these problems attacked the Captain's urinary system. He supposed that there could have been lots of reasons as to why the Captain become ill: from sexually transmitted diseases to perhaps a hernia in the body, but was more certain that the cause of the Captain's illness came from a urinary stone. He thought a stone, a calculi or "the gravel," was the cause because of the block of urination [17]. The fact that the patient reported that the calculi had burst from his scrotum served as further evidence. Smith further speculated that after the stone left his body, he started to heal, and his symptoms eased. A case of calculi such as the Captain's left a mark in Smith mind, and it was another source of fuel to encourage him to look for chemical solutions to treating the stone, so that he could further help his patients avoid suffering cases like this one.

Later, Smith saw two other major cases of urinary stones when he lived in Northwest South Carolina in the early 1800s. Both cases involve persons that will remain anonymous to history. In 1820, Smith published a write up of a case he had been party to in 1808 while living in Pendleton. He published "A Case of Dysuria" which was a description of a difficult case of urinary stones. A woman that Smith had known had not been able to produce urine for a long while, and no physician had been able to treat her case [44]. She had experienced difficulty with urination for about two years, but was in otherwise good health. Smith noted that she was [44] "stout and healthy appearance" but, because of the issues with urination, had been reduced "to a feeble and declining state: she had borne several children, but none since the commencement of her present complaint, nor could she assign any particular circumstance as giving origin to her malady." Smith, adhering to some conception of humoral theory, thought that commenting on the patient's constitution was a useful piece of information. Michael Foucault, a historian and a philosopher, argued that medicine went through a distinct transition during the nineteenth century from humoral theory-based medicine to a more specialized anatomical-based medicine [45]. Foucault expressed the difference in a simple analogy. He framed the difference as medicine transitioning from time consuming and slow medical consolations where the doctor inquired what foods the patient was eating, how was the weather, how was the wind, and how essentially did they feel. The transition to more anatomical and discrete understandings of disease then turned to quick visits that reduced the patient experience down to information, the doctor simply asking, where does it hurt? [45]:

> This new structure [of medical practice] is indicated—but not, of course, exhausted—by the minute but decisive change, whereby the question: "What is the matter with you?", with which the eighteenth-century dialogue between doctor and patient began (a dialogue possessing its own grammar and style), was replaced by the other question: "Where does it hurt?", in which we recognize the operation of the clinic and the principle of its entire discourse.

Smith also took a bold and heroic approach in this case that is also reminiscent of humoral medicine, in the similar manner of his teacher, Benjamin Rush. He took bold action to try and control the situation with the patient. Like the case above, Smith gave large doses of pain-treating medicines and purgatives [44]. He wanted to return the woman's fluids to a natural, flowing state and have her be able to pass urine. Smith, however, could not return her urine output to its normal levels. The patient permitted Smith more access into her private space and was allowed to use a bougie (which is a tube that allows urine to flow) and asked to give her an examination in her urinary organs, but Smith was refused. The patient later changed her mind, as she continued to suffer in pain. Smith was called to return and given permission, either from the patient or someone else—this is not clear, and he acts to attempt to end the blockage of urine [46]:

> Apprehending that there might be some mechanical obstruction, such as a small calculus in the urethra, which might require removal, I directed the patient to be laid upon a table as in the operation for the stone, and passing the fore finger of the left hand up the vagina, I introduced at the same time the end of a small probe into the orifice of the urethra. The probe stopped about midway of the canal, but the resistance to its passage did not indicate the obstruction to be caused by a solid body.

He continued to insert the probe into the patient's vagina, but the probe ran into an obstruction, so Smith added his finger to advance the probe. Smith eventually was able to get the probe into the patient's bladder.

During the nineteenth century, doctors believed that women had comparatively longer urethra than men and that they could avoid the surgery for the stone, and the doctor could simply extract it [47]. Smith was likely confident that he could extract the stone with his fingers, rather than risk surgery. However, when Smith added his finger to the patient's vagina to continue the probe into her body, urine started to emerge out of her body. In order to continue removing the urine, Smith added a larger bougie and continued to open the patient's urethra. The patient had some swelling, but most of the trapped urine was drained [46]. Smith left a reed in the patient in order to continue to remove all of the urine from her body.

Like the previously published cases, Smith mused about the larger causes of the trapped urine. He hypothesized that a urinary stone was likely the cause of the trapped urine, as he felt something press against his finger and then it traveled into the bladder. Smith could not say for sure, but he was certain that there was some type of blockage. He cited that the woman's constitution stabilized as she was able to have another child. But he wanted to communicate this case, like the previous case of the Captain, in order to inform other doctors who might be faced with the same case, and he hoped that this case might guide them in regards to interventions. The journal article was an important piece of information that circulated around health providers interested in treating urinary stones. Journal articles in medicine also became more important to not only medical communities, but scientific ones as well, and needed to be consumed by active members interested in problems of the body [48].

Published along with the case of the woman suffering from urinary stones was a case of a child suffering from stones as well. This was the case mentioned at the beginning of this chapter. Smith also deemed "A Case of Calculus in the Urethra of a Child Five Years Old" important for other doctors, as this case too provided insights into treating stones [1]. Since birth, the unnamed child had problems with urination, which were accompanied by pain and fever [1]. Like in the previous cases that Smith had published, he was called in when other physicians could offer no assistance. These narrative turns likely established his own credibility as an expert physician [49]. The child was suffering from problems not only in the urinary organs, but around them as well [1], "... a small hole was formed in the urinary canal, near to the anus, from which orifice the urine flowed out, and an abatement of the symptoms immediately took place." When Smith was getting the history of the child's problems from the boy's father, Smith focused on the problem of the hole where urine was coming out. Smith published his thought process in the article, theorizing that a stone being expressed from the patient's body caused the hole, as the stone probably originated from the patient's bladder, and found its way into the urinary organs by means of the ureters. The child's urine flow was blocked, this caused another hole to open, and the urine was expelled through this hole. Smith was able to examine the patients around two weeks after he hears about the case from the child's father. After the patient arrived in South Carolina, Smith was able to confirm his initial ideas. Like the case of the woman that was previously discussed, Smith examined the child and tried to find a stone [1]:

> ... [t]here was a fistulous opening the perinaeum, within a few lines of the verge of the anus; and I remarked that air seemed to come occasionally through this opening, which caused me to apprehend that there might be a communication betwixt the rectum and the bladder.

He noticed that the child also had swelling in the penis as the urine was backed up and, like the other cases, suspected that a stone was involved. As an adherent to humoral pathology, Smith wanted to unlock the urine and restore the fluids of the body to their natural balance, much like in other cases. Smith then preceded to make an opening to drain the buildup of urine. He then applied a dressing of lead (likely saturnine salts) in order to offer the wound, make by the lancet a type of ointment [50]. The area became less inflamed and pus ceased, and Smith then made the opening larger and was able to remove the stone with a pair of forceps, which the patient was able to deal with the pain of the removal [1]. The stone removed was larger, about the size of two ancient coins, according to Smith.

Two days later, after Smith was following up with the patient, he seemed to have recovered and improved. Like the woman, Smith deduced that he was successful because their constitution improved. Smith commented that the patient after the extraction [1] "… was become a stout and ruddy boy, able both to walk and run, which he never could do before." Smith was now intimately familiar with the problems that patients suffered from the stone. He moved his focus away from medical practice, and even the life of the planter, to engaging in chemical research to further understand the stone and hoped to find ways to prevent their formation.

4.5 Smith's Turn to Chemistry

During the nineteenth century, Smith started using chemical interventions in his medical practice. Around 1819, Smith wrote of the use of prussic acid in treating consumption in South Carolina [51]. Smith was working with another physician, James Davis, a physician that was originally from Maryland but moved to Columbia, South Carolina, to practice medicine. Davis had written much about an epidemic that attacked the throat in 1815. Davis was highly involved in the founding of the insane asylum on Bull Street in Columbia and was also interested in vaccination [52]. Though Smith turned to the practice of chemistry during the later part of his life, he returned to his role of physician during times of need, such as during this mysterious epidemic of "croup" that broke out in Columbia between 1815 and 1816 [53]. Smith tried to use a chemical remedy to treat consumption, in this case prussic acid. Consumption in the nineteenth century was likely a form of tuberculosis, and prussic acid was essentially cyanide. In the chemical literature, Smith had read of physicians like Dr. Francois Magendie, a Parisian doctor, who was also examining chemicals and their application to urinary stones [54, 55]. Smith was also in contact with Caribbean physicians who had migrated to the coastal city of Charleston [56, 57]. Felix Louis L'Herminier was originally from Guadeloupe and arrived in Charleston in 1815. He was able to produce the prussic acid for Smith's needs, but when Smith exhausted his supply, he was able to produce his own by learning how to do so in the work of Thomas Thomson [57, 58]. Smith wrote of the dangers of the medication, as it was a type of poison. He cautioned the reader that prussic acid was dangerous [51] "As to the nature of his substance, it is a most virulent poison, and in this respect you will recognise its analogy to some of our most effectual remedies." He then directed his reader to read one of the authorities about poisons and toxicology in the nineteenth century, Mathieu Orfila [59]. Smith did administer it to patients, knowing the danger, and gave them three drops, that was diluted with water, over the time of a day and a night. If the sufferer continued to have consumptive symptoms, they could take an additional five to seven drops in that time span. Smith did not know of anyone who had tried ten drops in twenty-four hours. Smith saw success in the cases as most of his cases recovered, which numbered around eight or nine patients, they all got better, save for one patient. He described the severity of the symptoms of the long-suffering patient [51], "… the distressing cough, copious expectoration, and wasting hectic were for weeks kept at bay, and the patient so much re-animated as to induce a hope of recovery; but this finally proved delusive." The prussic acid had helped ease the patient's suffering, so it was not wasted. However, some patients did complain about side effects, such as pain in the chest, expelling blood from their lungs, and some effects on the mind, but Smith stopped those patients from continuing the medication, and they returned to their normal state.

Throughout the decade of 1810–1820, Smith continued to publish about his interests in chemistry. In 1818, he analyzed water near Asheville, at the French Broad River. Chemists at the turn of the nineteenth century were highly interested to test their chemical skills in analyzing water [60]. Smith wanted to know what made up the mineral waters that were thought to heal people and be generally good for their health [61]. Smith was interested in researching the waters to see if they improved a person's health and looked at the temperature and general chemical makeup of the mineral waters. He found that the waters helped with diseases of the lungs and heart [62]. He noted that though the waters had healing properties, they could also provide ill health effects, like making the body's temperature rise too quickly. Temperature of the body was another aspect of humoral pathology that underwent revision in the nineteenth century [63–66]. As a quantitative chemist, Smith focused on the physical appearance of the water, how it tasted, how it smelled, and what type of bubbles it produced [62]. Smith, in his laboratory, added reagents to the water and recorded the reactions. He was looking for fixed air or carbonic acid in the water, as those two types of airs and acids helped people with stones and other health problems [62]. He even consulted users of the healing water baths that found them helpful, but their effects quickly ended. Smith recorded that the waters were useful for those suffering from [62] "... rheumatism, palsy, or loss of motion from other causes." Smith found that the more chemical research he engaged in, the more he hoped that he understood how substances affected the internal fluids of the body.

The last article that Smith published was his first comprehensive blending of both questions related to health and chemistry. The article "On the Application of the Medico-Chemistry to Calculous Affects" would be the apotheoses of both his research in chemical and urinary stones and was published after he died [67]. The editor of the *American Journal of Science and the Arts*, Benjamin Silliman, offered a brief obituary of Smith [68, 69]:

> This is the last communication for this Journal with which the editor was favoured by the respectable and estimable author of this memoir; it was transmitted a little before his death, but it has not been convenient to publish it before.

The article contained the excitement and optimism that Smith exuded about chemistry to the problems of diseases. Smith explained his exuberance about chemistry [69]:

> Although it must be confessed that a rash enthusiasm may have unwisely attempted to explain the mysteries of some Phenomena, that are observed in the living system, by the analogy of the results of the action of chemical agents upon dead matter, it must be granted that there are cases, in which the useful application of chemical knowledge is conspicuous.

Imitating the style of his dissertation, he paid homage to those chemists he perceived to be doing great work, but he was especially was interested in the work of William Hyde Wollaston and William Brande. Both men had been trained as physicians but gravitated to chemistry [70, 71]. Smith cited their work as inspiriting, especially their work examining urinary stones [69]:

> ... a light that is very cheering to the friends of science and humanity. We are no enabled to take a clear and satisfactory view of timidly groping in the blind paths of empiricism, we may walk bodily upon the highway of correct principles. This is the sure road, and if we are careful not to deviate from it, must gradually conduct us towards the attainments of our object.

Smith wanted to move beyond the anecdotal articles that he published earlier in his life and promoted the hope that chemistry would provide general knowledge that could help the physician in most cases, not just special cases of the stone. He was reading *The Edinburgh Review* and highlighted the good work that was being performed, such as the search for chemicals that would dissolve urinary stones. The search for chemicals that destroyed urinary stones was something that was of the upmost value to humanity, or as Smith punctuated its importance with a quotation from the *Edinburgh Review* that it was [67]: "... one of the noblest problems in practical chemistry, and among the best services that science could render to the healing art."

But Smith was skeptical that such a dissolving chemical could be found, or taken into the body, especially after the failure of famous chemists like Joseph Black. However, he was more likely to believe that certain chemicals could stop the development of stones in the body, such as in the work of William Grande and Sir Everard Home [72]. There was also hopes that better identification of stones would help to find the right chemical agents to help the patient. Smith was also excited about work done in researching the composition of urine. He thought that studying urine was important to advancing chemical understandings of stones as it was [73], "... of signal benefit and affords us a clue in the labyrinth, that would otherwise be impervious." Knowing what a patient's urine indicated could predict the formation of stones in the patient's body. Better knowledge of the chemicals of the urine could also indicate the type of stones that the patient was forming in their body and help the physician prepare treatment. Smith was advocating for a type of chemically informed uroscopy [74]. As a chemist, Smith was advocating for physicians and chemists to read the makeup of the chemicals in the urine, which they thought indicated the formation of a stone. Smith summed up this position regarding urine as having "premonitory symptoms of threatening dangers" and was hopeful if the chemist or physician was paying attention the patient, they could intervene in a very meaningful way and stop the formation of the stone [73]. Smith and his interested community of chemists theorized that the more acid the urine was, the more likely a stone was to form, which is somewhat analogous to today [75, 76].

Smith's optimism for chemistry in helping treat stones was abundant throughout the article. He was interested if acid in the urine caused stones, or if chemists detected excess salts, would be markers for stone formation. He used the notion of stopping the formation of the stone during its "germ," though Smith meant it in germination and not in the use of germ as a contagion [77]. He also thought that chemists might be able to change the fluids in the stomach and change the ratios of acids in the body to make meaningful change in stopping the formation of stones.

The intuition of the physician was still an important part of treating urinary stones, and not having the right type of judgment could be detrimental to the patient. Smith even caution that not knowing enough chemistry or having enough medical experience could make the patient's condition even worse. He strongly advised against those physicians who would try to use chemistry or treat the stone carelessly [77]:

> To this difference in the constitution of calculous matter it is owning that both the strong and the weak acids have sometimes been used with eminent benefit; and yet the indiscriminate prescription of acids would frequently produce the most serious injury.

Smith also commented on the trend of consuming soda as a panacea of health. He said that excess consumption could also cause stones. The editor of the journal, Benjamin Smith Barton, corrected Smith's comments about cautioning the use of magnesia, but Barton reminded the readers that Smith likely meant that this was only in the case of excess uric acid, as magnesia and other substances were good at flushing out excess acid out of the stomach.

Smith continued to push the notion of intuition and experience in treating stones, as he used his own case of the course of a stone as an example. In 1817, he felt that a stone was forming in his own body because of his lack of movement and the airs around him being a bit stale. By 1817, he was a chemist working at South Carolina College since 1812, and he claimed that the air in the new addition brick building on campus did not provide fresh air and the air was very cold, a signal of humoral theory as stale and bad airs were thought to contribute to disease [78]. Smith had pains and spasms, which he treated with a glass of wine. After asking his medical colleagues for advice, he reported that they thought that his internal liquids suffered from "acrimony" and his own body was forecasting the formation of a stone. He treated himself with the usual arsenal of purgatives that had been seen in the previous articles about stones that Smith published, and he had success in averting the stone. He framed his own symptoms for a coming stone as also indicative that humoral pathology was still a good theory in medicine. He commented on the state of theoretical medicine [79]:

> The extravagance of theorists, in almost any department of science, has sometimes carried them so far beyond the bounds of rational induction, as to involve in one common condemnation both truth and error; and this perhaps, has been the fate of the Humoral Pathology. Very lately this subject has been ably treated by Professor Cooper, of Philadelphia in his ingenious discourse upon the connection of chemistry with medicine, and in which it has

> been plainly shewn that the application of chemical science throw much light upon the reprobated doctrine.

Unfortunately, Smith does not further comment about why he perceived humoral theory as having a falling out in the medical community. He also perceived of it as being "discarded" and that chemists in South Carolina gave good chemical evidence of its value, as in his dissertation, he argued that chemists have found substances do not change in the body. He cited the work of a "L'Heiminier" (which was also spelled L'Herminier) as detected mercury phosphate that the child had consumed in their urine [79]. Smith did not see chemistry and humoral pathology as mutually exclusive, but one was a logical part of the other.

Smith also engaged in discussions of the latest chemical works, citing the work of Thomas Cooper (mentioned in the quotation above) and the work of Alexander Marcet. Smith had read a review of Marcet's comprehensive *An Essay on the Chemical History and Medical Treatment of Calculous Disorders.* Smith refereed to Marcet's statistical information about the frequency of urinary stones and what populations got urinary stones. Smith was interested in why it seemed that women suffered from the stone more than men. He was interested in Marcet continuing to pursue questions related to why certain populations seemed to suffer more from the stone than others. Unfortunately, this publication would be Smith's last publication, but he left lots of words of encouragements when he completed the manuscript in 1819.

4.6 Conclusion

Smith's practical interests in the chemistry of the stone also had a bearing on theories of medicine and theories of chemistry. The successes and optimism Smith had regarding the chemistry of urinary stones motivated him to pursue chemistry, but also advocate for the return of humoral theory. Smith's experiences treating difficult cases of the stone motivated him to find other solutions to improve patient treatment, and chemistry seemed to be the panacea he was looking for. Changes in chemistry that the Chemical Revolution brought on seemed to fuel a type of humoral pathology revolution as well. The importance of chemistry in the validation of humoral pathology would also continue in the work of other community members interested in chemistry and humoral pathology that Smith had read and interacted with, like Thomas Cooper. Cooper, as a radical in all things, advocated the return of humoral pathology, through the evidence found in chemistry, in even stronger terms than Smith.

References

1. Smith ED (1820) Case of calculus in the urethra of a child five years old. Phil J Med Phys Sci 1:149
2. Sharpe VA, Faden AI (1998) Medical harm: historical, conceptual, and ethical dimensions of iatrogenic illness. Cambridge University Press, Cambridge
3. Smith ED (1820) Case of calculus in the urethra of a child five years old. Phil J Med Phys Sci 1:150
4. Bynum WF, Lock S, Porter R (1992) Medical journals and medical knowledge: historical essays. Routledge, New York
5. Baldwin M (2015) Making nature: the history of a scientific journal. University of Chicago, Chicago
6. Rosenberg C (1979) The therapeutic revolution: medicine, meaning, and social change in nineteenth century America. Persp Bio Med 20:488
7. Rosenberg C (1979) The therapeutic revolution: medicine, meaning, and social change in nineteenth century America. Persp Bio Med 20:485–506
8. Whorton JC (1996) The purgative plan of James Hamilton, senior: therapeutic philosophy in the early nineteenth century. Pharm Hist 38:159–174
9. Warner J (1987) The therapeutic perspective: medical practice, knowledge, and identity in America, 1820–1885. Harvard University Press, Cambridge
10. Driggers E (2015) Boundary stones: morbid concretions and chemistry of early nineteenth century medicine. University of South Carolina
11. Waring IJ (1964) A history of medicine in South Carolina: 1670–1825. South Carolina Medical Association, Charleston, pp 315–316
12. Josiah Smith Letter Book, 1771–1784: https://finding-aids.lib.unc.edu/03018/
13. Hamilton JN, Rovner ER, Turner WR (2008) Urology in pre-civil war Charleston. J Urol 180:477–480
14. Unpublished manuscript file using information from "The Family of Edward North and Elizabeth Tucker." See Otto ON (1996) The Family of Captain Edward North and His Wife, Elizaneth Tucker: In Bermuda, America, England, Canada, British India, New Zealand, Australia, and South Africa
15. LaBrode M (1874) History of South Carolina College: from its incorporation, Dec. 19, 1801 to Dec. 19, 1865. Walker Evans & Cogswell Printers, Charleston
16. Friedman LJ, Shaffer A, Schaffer A (1979) History, politics, and health in early American thought: the case of David Ramsay. J Am Stud 13:37–56
17. Smith ED (1802) A singular case of hydrocele. Med Repos 5:134–136
18. Edgar W (1999) South Carolina: a history. University of South Carolina Press, Columbia
19. Driggers EA (2022) "… [T]he movement of a celestial system than a human invention:" Abram Blanding and bringing water to Columbia. Water Hist 14:21–40
20. LaBrode M (1874) History of South Carolina College: from its incorporation, Dec. 19, 1801 to Dec. 19, 1865. Walker Evans & Cogswell Printers, Charleston, pp 98–99
21. Bryan JM (1976) Architectural history of the South Carolina College, 1801–1855. University of South Carolina Press, Columbia, pp 59–61
22. Minutes of the Trustees, South Caroliniana Library, vol 1, pp 232–239, 245–249, 252–253, 262 and vol 2, pp 9–12, 50–51, 62–63, 68–69, 70–73, 79–80, 87–88, 90–91, 94–95, 98–99, 115–116, 119–120
23. Brown CM (1989) Benjamin Silliman: a life in the young republic. Princeton University Press, Princeton, pp 279, 283, 296, 302, 304–307, 294–296
24. LaBrode M (1874) History of South Carolina College: from its incorporation, Dec. 19, 1801 to Dec. 19, 1865. Walker Evans & Cogswell Printers, Charleston, pp 89–93
25. Smith ED, Silliman B (1819) Historical Society of Penn, American Scientists, Simon Gratz, MSS Folder 3 Box 7/25

26. LaBrode M (1874) History of South Carolina College: from its incorporation, Dec. 19, 1801 to Dec. 19, 1865. Walker Evans & Cogswell Printers, Charleston, p 91
27. Davis Z (1984) The return of martin guerre. Harvard University Press, Cambridge
28. Coyer M (2017) Literature and medicine in the nineteenth-century periodical press. Edinburgh University Press, Edinburgh
29. Smith ED (1805) Being an attempt to prove that certain substances are conveyed, unchanged, into circulation. In: Caldwell C (ed) Medical theses: selected from among the inaugural dissertations, vol 1. Thomas and William Bradford, Philadelphia, pp 229–255
30. Smith ED (1805) Being an attempt to prove that certain substances are conveyed, unchanged, into circulation. In: Caldwell C (ed) Medical theses: selected from among the inaugural dissertations, vol 1. Thomas and William Bradford, Philadelphia, p 231
31. Driggers EA (2021) "The voice of the stomach": the mind, hypochdriasis and theories of dyspepsia in the nineteenth century. Hist Pysc 32:58–99
32. Smith ED (1805) Being an attempt to prove that certain substances are conveyed, unchanged, into circulation. In: Caldwell C (ed) Medical theses: selected from among the inaugural dissertations, vol 1. Thomas and William Bradford, Philadelphia, p 232
33. Smith ED (1805) Being an attempt to prove that certain substances are conveyed, unchanged, into circulation. In: Caldwell C (ed) Medical theses: selected from among the inaugural dissertations, vol 1. Thomas and William Bradford, Philadelphia, p 234
34. Smith ED (1805) Being an attempt to prove that certain substances are conveyed, unchanged, into circulation. In: Caldwell C (ed) Medical theses: selected from among the inaugural dissertations, vol 1. Thomas and William Bradford, Philadelphia, p 247
35. Smith ED (1805) Being an attempt to prove that certain substances are conveyed, unchanged, into circulation. In: Caldwell C (ed) Medical theses: selected from among the inaugural dissertations, vol 1. Thomas and William Bradford, Philadelphia, pp 246–247
36. Maehle A-H (1999) Dissolving the stone: the search for lithontriptics, in drugs on trial: experimental pharmacology and therapeutic innovation in the eighteenth-century. Brill, Leiden, pp 55–125
37. Scheele P (1966) Joseph black-rediscoverer of fixed air. JAMA 196:362–363
38. Coley NG (1971) Animal chemist and urinary stone. Ambix 28:69–93
39. Coley NG (1988) Medical chemistry at guy's hospital. Ambix 35:155–168
40. Coley NG (1980) The laboratories of the royal institution in the nineteenth century. Ambix 27:173–203
41. Dobson M (1785) A medical commentary on fixed air. T. Cadell, London
42. Smith ED (1805) Being an attempt to prove that certain substances are conveyed, unchanged, into circulation. In: Caldwell C (ed) Medical theses: selected from among the inaugural dissertations, vol 1. Thomas and William Bradford, Philadelphia, p 248
43. Smith ED (1802) A singular case of hydrocele. Med Repos 5:135
44. Smith ED (1820) A Case of dysuria. Phil J Med Phys Sci 1:147–149
45. Foucault M (1975) The birth of the clinic: an archeology of medical perception (trans: Smith AMS). Vintage Books, New York, p xviii
46. Smith ED (1820) A case of dysuria. Phil J Med Phys Sci 1:148
47. Marcet A (1817) An essay on the chemical history and medical treatment of calculous disorders. Longman, Hurst, Rees, Orme, and Brown, London
48. Romano TM (2002) Making medicine scientific: John Burdon Sanderson and the culture of victorian science. Johns Hopkins Press, Baltimore
49. Rowland JC (1999) Heroic measures: illness, medicine, and narrative in nineteenth-century American fiction. PhD Dissertation, Yale University
50. Churchill TF (1808) The new practical family physician, or, improved domestic medicine guide. Albion Press, London, p 591
51. Smith ED (1820) On the use of prussic acid in consumption cases. Quart J Sci Lit Art 9:418–420
52. Waring IJ (1964) A history of medicine in South Carolina: 1670–1825. South Carolina Medical Association, Charleston, pp 206–208

53. Davis J (1817) Account of the winter epidemic of 1815–1816, as it appeared in the adjacent to Columbia, S. C. J Med Sci Lit 7:162–179
54. Sampo MA, Kyle RA (1987) Francois Magendie: early French physiologist. Mayo Clin Proc 62:412
55. Temkin O (1946) The philosophical background of Magendie's physiology. Bull Hist Med 20:10–35
56. Sanders AE, Anderson WD (1999) Natural history investigations in South Carolina: from colonial times to the present. South Carolina University Press, Columbia
57. Waring IJ (1964) A history of medicine in South Carolina: 1670–1825. South Carolina Medical Association, Charleston, p 254
58. Thomson T (1802) A system of chemistry in four volumes, vol 2. Bell & Bradfute, E. Bralfour, Edinburgh, p 224
59. Sánchez JRB, Nieto-Galan A (2006) Chemistry, medicine, and crime: Mateu J. B. Orfila (1787–1853) and his times. Science History Publications, New York
60. Hamlin C (1990) A science of impurity: water analysis in nineteenth century Britain. University of California Press, Berkley
61. Jennings ET (2006) Curing the colonizers: hydrotherapy, climatology, and French colonial spas. Duke University Press, Durham
62. Smith ED (1821) An account of the warm springs, in Buncombe County, State of North-Carolina. Am J Sci Art 3:117–125
63. Verwaal RE (2017) The nature of blood. Early Sci Med 22:271–300
64. Verwaal RE (2020) Bodily fluids, chemistry and medicine in the eighteenth-century Boerhaave School. Springer, New York
65. Chang H (2004) Inventing temperature: measurement and scientific progress. Oxford Press, New York
66. Arikha N (2007) Passions and tempers: a history of the humors. Ecco, New York
67. Smith ED (1821) On the application of medico-chemistry to calculous affections. Am J Sci Arts 3:300–310
68. Browning PE (1934) Benjamin Silliman and the beginnings of chemistry at Yale. J Chem Educ 11:170–174
69. Smith ED (1821) On the application of medico-chemistry to calculous affections. Am J Sci Arts 3:300. See the footnote for the memorial
70. Haigh E (1991) William Brande and the chemical education of medical students in French R, Wear A, British medicine in the age of reform. Routledge, New York
71. Usselman MC (2015) Pure intelligence: the life of William Hyde Wollaston. University of Chicago Press, Chicago
72. Lawrence S (1996) Charitable knowledge: hospital pupils and practitioners in eighteenth-century London. Cambridge University Press, Cambridge
73. Smith ED (1821) On the application of medico-chemistry to calculous affections. Am J Sci Arts 3:301
74. Stolberg M (2015) Uroscopy in early modern Europe (trans: Kennedy L, Unglaub L). Ashgate Publishing Limited, Surrey
75. Wiederkehr MR, Moe OW (2011) Uric acid nephrolithiasis: a system of metabolic disorder. Clin Rev Bone Miner Tetab 9:207–217
76. Moran ME (2014) Urolithiasis: a comprehensive history. Springer, New York
77. Smith ED (1821) On the application of medico-chemistry to calculous affections. Am J Sci Arts 3:302
78. Balzer W, Eleftheriadis A (1991) A reconstruction of the hippocratic humoral theory of health. J Gen Philos Sci 22:207–227
79. Smith ED (1821) On the application of medico-chemistry to calculous affections. Am J Sci Arts 3:306

5 Radicalism and Humoral Chemistry: Thomas Cooper's Atlantic Journeys

No address that Thomas Cooper (Fig. 5.1) gave to an audience was not passionate or memorable. The man that would later question whether South Carolina should stay inside the United States was during the early part of the nineteenth century asking questions concerning why medicine had left the practice of humoral medicine. While serving on the faculty of the University of Pennsylvania, he published a major argument concerning the relationship between chemistry and medicine, but he also strongly advocated his position regarding the importance of humoral pathology. He gave a speech on these ideas to his colleagues at the University of Pennsylvania and later published it. His speech was the basis of his awarding of a doctorate from the institution formerly known as the University of New York (now known as NYU) in 1818. Cooper had no formally awarded degree, though he had many years of higher education, as he often dropped out or was forced out of schools and organizations. Cooper's speech was later published as *A Discourse on the Connexion Between Chemistry and Medicine* in 1818. Regarding humoral theory, Cooper punctuated his defense by stating that [1],

> Professors and practitioners of medicine, in every part of Europe, are now alive to the claims of chemistry, too imperious in its present improved state to be neglected. The time has arrived, when, however reluctantly, we must retrace our steps: nor is it difficult to shew that even the humoural pathology, stands upon much higher ground than those who smile at the application of chemistry to medicine, are willing to allow, or able to deny.

Cooper wanted to advocate the importance of chemistry to doctors, but also how important chemistry was to humoral theory. In this speech, Cooper was building on the ideas of other defenders of humoral theory, such as Benjamin Rush and William Cullen. He had call backs to like-minded doctors like Rush [1]:

> Such also was the case with the medical school of Philadelphia, the leading feature of Dr. Rush's theory being, that pathology is reducible to the morbid action of the living solids.

E. A. Driggers, *Early Nineteenth Century Chemistry and the Analysis of Urinary Stones*, Perspectives on the History of Chemistry,
https://doi.org/10.1007/978-3-031-34973-7_5

Fig. 5.1 Thomas Cooper M.D. (National Library of Medicine)

The humoural pathology, with its cacochymia, lentors, fermentations, and spiculae, could no longer be supported by the imperfect chemistry of the day.

Rush's humoral ideas were also discussed in the context of chemistry, as was presented in an earlier chapter in this book. Cooper wanted doctors and chemists to understand the remaining value of humoral theory as all-inclusive idea in chemistry and medicine.

Most historians associate Cooper with proto-Southern nationalism or the first attempt of South Carolina to succeed from the United States. Cooper seemed to take on the role of rebel rouser, as he antagonized monarchists in England, his home nation, in the wake of the French Revolution [2]. Cooper, who also served as a magistrate in the American judicial system, found himself on the wrong end of the law when he was charged with sedition against President John Adams [2, 3]. William Freehling, a historian of the antebellum period in American history, nicknamed him the [3] "Professor of Revolution." Cooper stoked controversy wherever he went, including his time at the president of South Carolina College after his stint at the University of Pennsylvania. Cooper not only irritated the conservative religious community with his ideas about geology and an older earth, he participated in the Nullification controversy, which occurred in South Carolina, where the state government thought that they could nullify federal law and state law was supreme [4, 5]. Freehling described Cooper down to his temperament [6], "he had come to crusty old Carolina, apparently the wrong place for a reckless hothead, at the right time, when the tariff and colonization crises were simultaneously occurring." Eventually, Cooper was fired from South Carolina College because of all the controversies and person enemies he had developed in the city [7]. Cooper was also a practitioner and defender of slavery [8].

Though Cooper created many enemies and most accounts describe his personality as abrasive, he did have some lasting friendships in the scientific community, such as that with Joseph Priestly, an active chemist and political dissident in the nineteenth century [9]. He also was a correspondent of Thomas Jefferson, and Jefferson tried to recruit him to be the chair of the chemistry department at the newly created University of Virginia [10]. However, Cooper received admonishments in parliament from Edmund Burke as well [11].

Cooper did practice medicine, but without a degree, and he practiced medicine prior to his doctoral degree from the University of New York. He felt that charging for his medical services was beneath his stature as a gentleman [12]. Cooper saw some of the worst parts of the yellow fever epidemic in the late eighteenth century, the same one as Benjamin Rush, who saw so many patients die. Cooper and other physicians fit the ideas of Charles Rosenberg, where they wanted to improve public health to prevent additional outbreaks of fevers [13, 14]. These physicians wanted to help the body restore itself and called for [13], "clean streets, airy apartments, [and] a pure supply of water, were certain safeguards against epidemic disease."

In Cooper's speech, he makes an overarching argument that controlling the body's fluids prevented or mitigated disease, much like many of the characters that are in this book. And problems like calculi, inflammation, and other diseases really originated from the internal fluids experiencing some type of blockage. Cooper both argued against panaceas, but presented one in his own works.

An important point of analysis that historians of chemists have not fully utilized is the notion of historical rhetoric. Historians can get a lot of information not only from the ideas that Cooper was discussing, but how he was presenting them [15]. Scholars of European medicine have notable success [16]. This chapter not only looks at Cooper's chemical and medical ideas surrounding humoral pathology, but how he expressed his support for them.

5.1 Thomas Cooper's Education and Background

Thomas Cooper's biography is incredibly interesting, and it is no surprise that one of the premier biographers of the twentieth century had one of his earliest publications on the life of Thomas Cooper [2].[1] Dumas Malone would go on to write a pivotal biograph of Thomas Jefferson, and it would win several awards

[1] See https://archives.upenn.edu/exhibits/penn-people/biography/thomas-cooper/ for an excellent biography of Cooper.

[17]. More works in regards to the history of philosophy, science, food studies, and slavery studies need to be written using the life of Cooper [8].

Cooper's backstory starts in industrialized Manchester. Thomas Cooper was born to a brick manufacturer. Biographers like Dumas Malone wrote about Cooper's early interest in industrial chemistry, perhaps originating from the family business. However, some biographers simply ascribe his origins as aristocratic [2]. The original article that Malone and other biographers of Cooper site was written in 1920 [18], which attributes a small fortune to Cooper as [18], "His father died about 1789, as which time he was owner of a tract of land containing brick-earth, comprising about forty acres, from which he derived in that year an income of nearly £500." Most biographies agree that Cooper was born around 1759, and he traveled to university on the family's good fortunes. Though Cooper likely had earned a degree in Oxford, he left the University in 1779 because of a disagreement of religious consciousness as he did not want to agree to the Thirty-Nine Articles, which summarized the major faith ideas of the Anglican Church in England [19]. However, Malone noted that this story had a family lore to it, so no one knows for certain why he left. But the later religious radicalism of Cooper seems to track with this idea [20, 21]. Cooper later commented on his frustration with the higher education and how he skirted the requirements of his exams at Oxford [22], "I passed well enough by construing a page in Horace, another in Demosthenes, and another in the easy Greek of Euripides, together with a few answers in Euclid, in logic, and some other trifling branch of education. This will not do there now." After marriage to Alice Greenwood, the child of a prominent shipping magnate, he turned to the practice of medicine, while having four children with Greenwood, though other biographers noted that there were five children [2, 18]. Cooper eventually followed the wishes of his family and practiced law, but cooper was able to pursue medical and anatomic education London and got further medical education at Middlesex Hospital [23]. The rumor of Thomas Cooper practicing medicine but never taking any fees originated in the 1920 article [23], "From these different sources he secured a medical knowledge which qualified him to practice among his neighbors and friends, as he was accustomed to do wherever he lived, though never accepting any of the frees offered because he considered it unprofessional in a lawyer to do so."

Cooper had a lot of interest in chemistry throughout his life. After finding work at a business firm in the name of Baker, Teasdale, Bridges, and then Cooper, as they were "calico-printers" near the town of Manchester. It was during that time that Cooper published a work on industrial chemistry published later in 1815 [24]. Cooper likely learned about chemistry at Oxford, but there are no good sources that focus specifically on his chemical education, though there is an article on the historical contributions of chemistry by Cooper [25, 26].

Cooper turned to the practice of law in 1787 and was in the law profession in Manchester for around three years. While in legal practice, he wrote against the slave trade [27–29]. Cooper became a promoter of the ideas of the French Revolution, held religious ideas that were a bit out of the mainstream, and his antimonarchy political stances necessitated him leaving the United Kingdom and coming to America with his colleague Joseph Priestly in 1793 [30]. He returned to England and publish a travel account of his time in America and sent it home to an interested audience in the England, that eagerly read his travel writings [2, 31]. Cooper and Priestly shared a close relationship because of their shared interests in politics, religion, and chemistry. Cooper later edited the writings of Priestly [32]. Cooper landed a position on the bench (judgeship) in 1804 in Philadelphia because of his connections with Thomas Jefferson, whom he had been corresponding with regarding science and also politics [2, 33]. Cooper had likely garnered Jefferson's attention because of his sharp criticisms in the newspaper (*Sunbury and Northumberland Gazette*) against the then president John Adams, whom Jefferson despised at the time.[2] Cooper was charged with sedition for disparaging Adams and sent to jail for six months with a fine of four hundred dollars [2, 34–36]. Then Cooper would later lose his judgeship after losing political favor in Pennsylvania [2]. During the political fallout and tribulations of the early nineteenth century, Thomas Cooper continued to read and educate himself regarding chemistry.

Though he lost favor with the Republican political establishment in Pennsylvania, he was able to secure an academic position at the college in Carlisle, which is now called Dickson College. Cooper took up the position in 1811 and served as the chair of chemistry at that institution [12]. Cooper's notoriety in chemistry had spread, and other chemists like Benjamin Rush supported him taking up the position at Carlisle [37]. Though there was a lot of support for Cooper coming to the institution, because of his personality, there became little support for him remaining in the college, and he left by 1815. In 1815, however, Cooper had secured another teaching position in chemistry in Philadelphia at the University of Pennsylvania [25]. Prior to traveling to Philadelphia, Cooper had been appointed chair of chemistry at the University of Virginia by Thomas Jefferson, but could not secure the position ultimately because of the objections of the religious establishment in Virginia, and the University did not open for classes until 1825 [38, 39]. In the meantime, Cooper worked at the University of Pennsylvania from 1815–1818. Cooper left the institution after he did not receive an additional role in the Medical Department and another chemist, Robert Hare, secured the position [40].

After dismal results in the universities of Pennsylvania, Cooper migrated South and found a job at the South Carolina College replacing the previous chemistry professor Edward Darrell Smith in 1820. The president of South Carolina College, Johnathan Maxcy, died in 1820, and Thomas Cooper became the president of the institution in 1821 [41–43]. While serving as the college president, Cooper continued to remain active in philosophy, science, chemistry, religion, and politics. Between his time at the University of Pennsylvania and South Carolina College, Cooper published *A Discourse of the Connexion Between Chemistry and Medicine*. The work is particularly important because it gives tremendous insight into the

[2] The National Archives (USA) produced an educational lesson about the newspaper sources that were used to charge Thomas Cooper with sedition: https://www.docsteach.org/documents/document/newspaper-broadside-cooper.

nature of chemistry and humoral theory as nineteenth century thinkers like Cooper thought about disease processes.

5.2 Cooper's Perception of Humoral Pathology

Cooper offered a historical interpretation of humoral theory to the point of his own lifetime. As a comprehensive theory of medicine, humoral theory lost its explanatory power because the chemistry of humoral theory did not keep up. Chemistry, Cooper argued, was to blame [44], "the humoral pathology, with its cacochymia, lentors, fermentations, and spiculae, could no longer be supported by the imperfect chemistry of the day." Chemistry, in Cooper's time, had a bit of a rough relationship with medicine, and some doctors did not think that there was any value in chemistry for treating patients. Scholars have noted that early chemical education started in medical schools and in medical courses before moving into their own departments.[3] There was an idea in medicine at the time that [44], "Chemiae in medicina, fere' nullus est usus." This roughly translate to Chemistry in medicine, and there is hardly any use. The original saying perhaps comes from Stahl, a chemist in the early modern period [45].

Though the chemistry had not been developed to support humoral theory, doctors now have a better command of chemistry, and according to Cooper, believe that chemistry is an important tool in medicine. With the improvement of chemistry, he argued, doctors need to go back to older theories of medicine. Cooper directs that [44], "The time has arrived, when, however reluctantly, we must retrace our steps: nor is it difficult to shew that even the humoral pathology, stands upon much higher ground than those who smile at the application of chemistry to medicine, are willing to allow, or able to deny."

Like other theorist in this book, Cooper believed that diseases processes occurred in the fluids that were between the solid parts of the body. Though the solid parts of the body, which include muscle and bones, were outside of the scope of chemical theories, the fluids did adhere to a chemical explanation of diseases in the body. Cooper also shared the idea of the morbid sympathies in the body, where problems in one parts of the body could spread to others [46]. He was arguing against doctors who thought the truth in medicine was explaining all the morbid sympathies, which Cooper countered as saying that it would be unreasonable for humoral theory to have to explain all of those phenomena in the body [44]. Cooper proposed that humoral theory could account for the idea of sympathy in the body [47]. He did not think it was necessary to choose between the two

[3] For more information on the history of chemical education in universities, including its transition from the medical curriculum to its own discipline see Newell LC (1932) Chemical Education in America from the Earliest Days to 1820. J. Chem. Educ 9:677–695 and Browne CA (1932) The History of Chemical Education in America Between the Years 1820 and 1870. J. Chem. Educ. 9:696–728.

theories [47–49], "...there is too much truth both in the one doctrine and the other, to reject either altogether."

Cooper thought that any single explanation of illness was not useful. He used the example of gout to prove his point. Doctors believed that gout was caused by a severe problem related to the body's internal fluids, but they were surprised that interventions like breathing the vein or purgatives did not make a difference and the patient did not improve. Theories of gout explained the condition through having too much acid in the body, which caused the body to become inflamed. Other theories thought that stones produced gout, or there was too much of a uric acid (or lithic coming from the word stone) caused blockages of the body because of excessive consumption of drinks that were too acidic. But other doctors thought that some people were at risk of getting the disease because of their constitution. Cooper liked the idea that gout came from bowels that were overworked. And he recommended balance in diet or restriction of foods.

Cooper's passion for his ideas came through the page to the reader. He was in the position that no single causes of disease was correct and disease was caused by many different factors. It was likely that he focused on multifold reasons for causes for disease in order to make room for humoral theory. He reminded the reader that [50]:

> intolerance is the bane of improvement. Intolerance in politics, intolerance in religion, intolerance in medicine. Those who deem themselves entitled to the patent right of the system, will admit to no competitor and no sharer. Everything must be done and explained by the sole instrumentality of their exclusive method.

Cooper wanted a more flexible theory of medicine. In particular, he wanted medicine to move away from this idea of "exclusiveness" and wanted doctors to be cautious about novel theories of medicine because they often did not work out [50]. He was also wary of the idea of a panacea [50], "new theories explain everything: new remedies are panaceas: hence the propensity in all of us, to reject what is useful from its occasional alliance with unfounded pretension." Though Cooper would accept any medical idea that seemed to work in practice, he wanted to make room for the return of humoral ideas to medical practice [50], "But the humoural pathology, which admits that the causes of disease may exist in the fluids as well as in the solids of the body, stands, as I suppose, upon ground too firm to be shaken." And though Cooper was tolerant of other medical theories, he was making this speech to really advocate for humoral theory, and he did so forcefully through the use of rhetoric.

5.3 Cooper's Work to Validate Humoral Theory

In order to excite his audience and readers, Cooper cited the work of chemical authorities. He referred to Everard Home and John Hunter, surgeons who were also interested in chemistry [51]. But Cooper tried to logically probe the importance of chemistry to humoral pathology through twelve to thirteen points [52]. Cooper uses the idea that the fluids of the body causing disease and basically casts that idea as essentially the new humors of the body. As the reader recalls, the classical humors of the body included blood, phlegm, black bile, and yellow bile [53]. The general fluids of the body are basically humors to doctors and chemists like Thomas Cooper.

The fluids of the body were crucial to transforming goods into blood, and food was transformed into other vital fluids. Chyle produced blood as the pancreas and the liver were important in changing food into bodily fluids [54]. Chemicals also worked to affect the fluids of the body, such as salts and acids. Gasses, electricity, and other stimulants affected the fluids of the body. The body actively disintegrated foods, acids, salts, and other substances and transformed them. Therefore, reading each fluid of the body illustrated the amount of substances the body was taking in, much like a complete blood count (CBC) panel that doctors subject patients to today. Sweat and expelled urine indicated the nature of all of the body's fluid, especially its character. Whether the fluids were acidic or basic in their nature informed the doctor about how the patient's body was performing or whether these fluids needed adjusting. Referring to the work of Edward Darrell Smith, among other chemists, that was true that fluids traveled in and out of the body unchanged. Poisonous substances also affected the fluids and caused disease processes. Cooper cited contagion diseases like small pox, as well as cancer, that changed the solid parts of the body through its fluids. Cooper summarized the changing nature of fluids in the body by referring to a saying from chemistry *corpora non agunt nisi soluta.* This saying meant [55] "(substances do not interact unless dissolved)," to which Cooper added that water was the most important substance in chemical reactions. He believed that he could prove that the fluids contributed to the actions of the body and believed that he could provide [56], "…an outline, however, that sufficiently proves the fluids of the body may be diseases themselves, and may produce morbid action in the solids."

Action defined the life processes in the body, according to Cooper, and the fluids of the body were much alive, and many philosophers like Cooper thought of living systems having a life force. And perhaps changing the fluids could change the actions in the body [57]. Cooper cited Newton and put Newton's ideas into his own words in respect to the making a change in one part of the body, the fluids, having an affect on another, the solids [56], "It is a universal law that action and reaction are equal and contrary." Cooper, like many chemists and doctors, believed that there were laws that controlled the body, much like those in the *Principia*, Newton's major work that attempted to explicate the laws of the universe [58].

Cooper saw the body, in many ways, also in chaos. The internal fluids were chaotic and being converted into new substances, which prevented affinity in the body [56]. As a chemist, Cooper likely saw the body in chemical terms. Affinity in chemistry is when substance come together and form a new substance, it is essentially the force that holds everything together [59–61]. He speculated that electricity in the body prevented that processes, but disease occurred when affinity happened the body and the fluids come together [56]. Fluids not governed by the body and achieving affinity with other fluids or parts of the body produce "morbi" which Cooper discussed in a type of disease-spreading agent or constituent part [62]. In a quick dismissal to objections to his theories, he gave an aside that only those persons steeped in high chemical knowledge, such as the process of affinity, will understand his arguments. But then cooper turned around and continued to embrace the virtues of the importance of chemistry to medicine.

Cooper was making a very rhetorical argument about the importance of chemistry to medicine, as he wanted to make the argument like he was giving a public disputation [63–65]. Cooper had a reputation as a forceful orator, as he strongly supported his controversial religious ideals [66]. He matched that same energy with writing about chemistry, politics, and even medicine. Cooper was also friends with other chemists who was not afraid to share his opinion, Joseph Priestly. Priestly debated with Lavoisier on the nature of combustion. And though some scholars argue that Joseph Priestly should be credited with the discovery oxygen, Priestley also supported phlogiston theory. Priestly published on the gas first, but he might not have been the first philosopher to make the discovery. Phlogiston theory is the idea that substances combusted or burn by release phlogiston, an invisible substance that was released in the fire when a substance was burning [67].

Discussing Priestley's language and arguments about oxygen reveal the language of science in the nineteenth century according to Nan Johnson. Johnson contextualized the work of Priestley [68]:

> "Joseph Priestley's *Course of Lectures on Oratory and Criticism* (1777) relies on the assumption that 'two sources of the principles of human nature and pleasures of the imagination…explain the efficacy of rhetorical devices…the association of simple ideas [and] a moderate exertion of the facilities" (xliv).

Cooper was using the same types of rhetoric to argue for the place of humoral medicine in chemistry and in medicine generally in the nineteenth century. Priestley is credited with setting the standard for arguments in the nineteenth century science by scholars like Johnson. Because Cooper and Priestley were close friends, it makes sense that Cooper's language and rhetoric follow the same patterns and inspirations.

In the *Connexion Between Chemistry and Medicine,* Cooper used lots of case studies and clear illustrations to make his points. Cooper said that chemistry was the key to most subjects, even beyond medicine [62]: "…pathology, semeiology and therapeutics: poisons: mineral material medica: vegetables materia medica: the adulteration of medicines: pharmacy: and prescription." He also used vivid descriptions of experimentations and used the work of Priestley as an example [62]:

> …shewing the disappearance of oxygen when exposed to venous blood inclosed [sic] in a thin bladder, and the florid colour thus produced. At present, however, we know tolerable well, that the air inspired is little changed in quantity: that the oxygen is converted into carbonic acid: that a small quantity of additional moisture is contained in the expired air; and that it is thus fitted for stimulating the left ventricle of the heart. At this period, and

not sooner, does the chyle assume the properties of the blood; whether a repetition of this process be necessary to furnish a full share of vitality to the chyle is not yet fully known.

Every example that Cooper gave was a subtle reminder about the validity of humoral theory. Chyle was an important fluid that provides life giving substance to humans, and it involved one of the most important organs in humoral theory, the liver, where chyle was energized [54].

Cooper wanted to continue to emphasize how chemistry could reveal overall health in the body. Fluids told a story about disease in the body, and chemistry was the compass to lead chemist and physicians to diseases. Using chemical analysis, Cooper promoted that chemist and physicians would understand the chemistry of life, such as the processes of breathing [69, 70]. Chemistry was touted to reveal secrets of other biological subjects and studies of the body [69], "Whatever is known of these processes, we owe to chemistry, and to chemistry exclusively." Cooper continued his emphasis on the importance of chemistry to reveal diseases processes in the fluids of the body.

Chemistry revealed diseases processes in bodily fluids like urine, bile, and especially blood. Chemist could break blood into its constitute parts, such as albumen, as well as the fibrin, and doctors had found that chyle was circulating into the blood that eventually traveled to the heart, sending vital energy throughout the body [71]. He argued that blood transported oxygen because of the physical color of the blood, and it was also responsible for removing toxic substances from the body. He cited the work of one of the most famous turn of the nineteenth century Jöns Jacob Berzelius.

Berzelius was interested in animal chemistry, among many types of chemical research, and worked with Alexander Marcet. Marcet will be discussed later in this book. Both men were also interested in how the blood functioned in the body as another important part of the fluid network of the body. Cooper then ended his discussion of blood by arguing that there is some type of affinity processes that occur in the blood because some gases or other substance must covert some of the body's blood into gelatin, as gelatin was found in the solid parts of the body, such as the skin, and he thought that blood transported that material to the solid parts of the body.

Cooper then turned his attention to discussing the bile found in the body. Bile was one of the original four humors. He articulated as physiological explanation of how bile worked in the body. Bile brought important power through the "vena portaum" which was a vein of some size that communicated digested food to the solid parts of the body. Bile's color indicated diseases processes in the body, as it was naturally yellow, and green, or even lacked color when the body was sick. Through his experience and the medical-chemical literature, Cooper noted that consumed food changed the bile, such as if the food had a high amount of acid in it. He prompted that more study and careful chemical analysis could further reveal markers for disease in the body.

The last bodily fluid that Cooper presented in depth was urine, and he believed that there was a lot of potential in the analysis of urine. Cooper promoted the promise of the analysis of urine to the reader [72], "to chemistry, we owe our knowledge of the general composition of healthy urine: and to chemistry, we owe all that is known of the variations that take place when urine is secreted, either from animal fluids that are morbid stimuli, or by morbid action of the kidneys from healthy fluids." Like many of the medico-chemists presented in this book so far, Cooper illustrated the many components of urine and touted chemistry providing the tools to read the health of the body.

Cooper also tied other diseases to the fluids of the body, such as the piles (hemorrhoids) and hysteria, two conditions that perplexed doctors. He hypothesized that the body's fluids becoming too acidic created the conditions the body through irritation. Chemistry, Cooper optimistically proclaimed, was a tool that doctors could use to manipulate the body's fluids and reverse diseases in the body, like hysteria. Hysteria was located in the stomach and was spread through irritations in the "primae vae" or in the uterus. Though throughout history conceptions of hysteria were based on the women's uterus, but Cooper firmly argued that most diseases originated in the stomach and cited the work of Jean Fernel Fernelius [73, 74]. Cooper believed in Fernelius work and cited Fernelius's motto [75], "...all morbid concoction and impurity of the humours of the body, proceeds either from a diseased affection of the stomach and viscera, or from a gross and faulty diet." He was convinced that all diseases had their origins with too much consumed acids that affected the stomach somehow, where the irritation spread throughout the body. Cooper also explained mental illnesses the same way, through the spread of irritated fluids in the body [76].

5.4 Weather, Disease, and Humoral Theory

Cooper's ideas about disease were also inspired by the Hippocratic ideas about the link between diseases and weather, especially its effects on the fluids of the body. Hippocrates, a doctor in classical Greece, has become the originator of what most historians associate with humoral theory, even though there were contributors to these medical ideas from China and the Middle East. Hippocrates was not only just one doctor, but the Canon of Hippocrates was also made up of students and others writing in the name of Hippocrates [77–81]. Ideas in humoral theory not only emphasize the need to balance the fluids of the body to prevent disease, but doctors versed in humoral theory had to understand how diets, lifestyle, and the weather affected the internal processes of the body. In the Hippocratic writings, the essay "Airs, Waters, and Places" details how weather was involved with disease [82], "Whoever would study medicine aright must learn of the following subjects. First, he must consider the effect of each of the seasons of the year and the differences between them. Secondly, he must study the warm and the cold winds, both those which are common to every country and those peculiar to a particular locality. Lastly, the effect of water on the health must not be forgotten." Doctors needed to understand diseases holistically, including the effects of weather, or would face negative outcomes for their patients [82]:

> A physician who understands them well, or at least as well as he can, could not fail to observe what diseases are important in a given locality as well as the nature of the inhabitants in general, when he first comes into a district which was unfamiliar to him...Being familiar with the progress of the seasons and the dates of rising and setting of the starts, he could foretell the progress of the year. Thus, he would know what changes to expect in the weather and not only would he enjoy good health himself for the most part but he would be very successful in the practice of medicine.

Cooper expressed similar ideas about disease, the fluids of the body, and the weather in his writings emphasizing the causes of disease [82, 83], "...since the changes of the seasons produce changes in diseases." He also thought that weather reports and study could explain the breakout of disease. Weather was a very important aspect to personal health during the nineteenth century, and many doctors believed that specialized fevers broke out during periods of distinct weather, and some philosophers even hypothesized that plague and other contagion broke out in specific weather patterns [84–87].

Cooper's thinking about nature represented most nineteenth century conceptions of pure water and land, that running water was clean and lands near swamps were disease causing areas [88]. Swamps and weather perhaps contributed to the production of miasmas in the body, such as in the liver. The liver removed pectant and other noxious humors from the body, but if it reached a heightened state, there was a systemic problem in the body, often resulting in a fever. The bile would often change colors to indicate a diseased state. Chemist paying attention to the body could spot problems with the liver with the changes of fluid, such as distinct changes in the sweat that produced a new fluid that resembled resin.

Readers need to remember that fevers were not a symptom of disease but a disease of themselves in nineteenth century medicine [89]. And some fevers were deadlier than others, such as bilious fever. Cooper, as a chemist, offered a chemical understanding of fevers that were caused by acid. He cited examples from New Orleans where yellow fever suffers had a buildup of acid, and the French thought that by teeth chattering in fever was caused by excess acid in the body from a fever [90]. The acids from fevers are what destroy the body as the acid [91], "...acts on the stomach in yellow fever, disorganizing and destroying the coasts of that organ, and converting them into the dead matter of black vomit." And by knowing that acids are what drive the destructive nature of fevers, Cooper thought that he might have a way to intervene and speculated that chemistry could stop the acid in the priae viae in the body, as it ran from the mouth to the anus, and chemistry could end the symptoms of the deadly acid [91] and stop the, "...train of symptoms that a recurrence to the known facts of chemical affinity will alone be competent to combat." In confusing language, at one point Cooper blamed chemical affinity for the causes of disease, while in subsequent pages, touted it as a way to end the excess acid in the body.

Excess acid, Cooper speculated, was to blame for other conditions, such as trouble in the stomach known as "dyspepsia." Summarizing his ideas about fevers, fevers could be categorized as stomach troubles that affected other parts of the body, such as the liver, digestive track, or even the pancreas. He called for doctors to use the same medications that treated stomach pain for doctors to use in treating fevers, such as soda. Acid produced in fevered state caused the body to also respond when it encountered this diseased fluid. Diseased acids of the body, Cooper pointed the blame, at causing other conditions such as cholera or even dispensatories, or the patient suffering diarrhea. But the chemist could now intervene in these conditions to change the acid in the body and provide medications that could be used as astringents or combat acids like terra japonica, but could also use kino [92]. Hemorrhoids and the piles, as well as diseases of the liver, which included chlorosis, were also thought to be caused by excessive acid and could be treated now by the chemist, as the disease bile could be fixed. As a physician educated in the classics, he quoted from the work of Jean Fernel (Fernelisu), which argued that all disease came from diseased humors or excessive food consumption [75, 93], "*Omnis enim cacochymia, et humorum impuritas, aut ex vitiosd viscerum affectioney aut ex improba vivendi ratione, raro aids ex eausis, projiciscitur*," and Cooper deciphered the French into the English as, "That is, all morbid concoction and impurity of the humours of the body, proceeds either from a diseased affection of the stomach and viscera, or from a gross and faulty diet." Illness could originate from the stomach and intestines for many reasons, and spasms in these organs could lead to whole body irritations, including the production of miasmas that ultimately produced contagions [94, 95]. Other conditions could be explained through whole body irritation, resulting from excess acid, and these conditions included urinary stones, diabetes, poisoning, cases of gout, and other conditions. He was convinced that gout was caused by too much acid in the body, though other factors like family history might be important, but it was the spasm in the body (he called morbid action) that was truly to blame.

Other substances, such as alcoholic beverages, also contributed to the production of more acid in the body, including lithic and uric acid, which contributed to gout. Also, that bodily acid formed urinary stones as well. Though chalk stones were destroyed by bodily acid, and these stones also could be used by medicochemists to treat stomach diseases. Chemistry revealed the uses of stones, and Cooper promoted to the reader that [96]:

> all that we know of the composition and formation of these productions of gout and gravel, we owe to chemistry alone: and, cathartics excepted, all the remedies hitherto suggested, have been furnished by chemistry, on chemical considerations: the same may be said not merely of the remedies but of the prophylactics also. In fact, without chemistry, nothing would have been known of the theory or the cure of gout, stone, and gravel; although, as I allow, other considerations may enter into our view, as gastritis, hepatitis, and nephritis, whether owing to original affection, to metastasis, or to sympathy.

It was clear to Cooper that chemistry was useful to medicine, but it was not accepted by all doctors. For instance, Nathaniel Chapman admitted that chemistry might be useful at studying stones, but could not accurately be used to identify stones [97]. Chapman openly chastised physicians like Thomas Cooper, batting down any contribution of the chemistry to medicine [98]: "As applied here, the chemical resolution cannot be trusted. Experiments have fully demonstrated, that articles widely discrepant in their genera nature, as aliments and medicine, the most salutary good, and the rankest poison, exhibit, on analysis, nearly the same results." He provided further caution to the reader [98], "This indeed holds so generally true, that the virus of the viper, and the mildest mucilage, constitute no exception; as, when resolved into their elementary principles, they are precisely analogous." Chapman was a major medical figure in Philadelphia and apparent straw man for Cooper [99, 100]. Cooper then channeled the work of another medico-chemist, Alexander Marcet, who will appear in the next chapter; Cooper used Marcet's chemical work in studying urinary stones as a clear answer to the problems of identification that Chapman made in his skepticism of the usefulness of chemistry in medicine.

Chemical knowledge, to Cooper, was a fulcrum that could divide amateur or irregular practitioners from those physicians that had proper scientific knowledge [101]. He admonished practitioners who did not have chemical knowledge [102], "owing to the want of chemical as well as pathological knowledge generally, patients in this disorder have been from time immemorial the prey of quacks out of the profession and empirics within it." Also, he commented on the dangerous nature of patented, or homemade medicines that were often sold by less than reputable individuals [102], "...useless in the hands of skilful men, and most dangerous in the hands of common men: the best opinions and observations agree that, in every case whatever, if they shorten the paroxysm, they lengthen the disease." Doctors, according to Cooper, are the only individuals who know the body well enough to safely prescribe medicines. Being versed in chemistry made the physician a safe operator treating and identifying the stones of the body, even against those healthcare providers that might have more experience.

Diabetes was another disease that could be treated by the chemical versed physician. Cooper noted that there was a chemical debate regarding the causes of diabetes, whether the sugar appeared in the blood. The debate was between two major medical chemists, William Hyde Wollaston and Alexander Marcet. The debate will be more closely examined in the next chapter on Alexander Marcet.

Cooper was optimistic that the relationship between chemistry and medicine was growing ever closer. Medical and scientific journals at the time of Cooper's life often ran medical cases alongside informational articles about chemistry. Cooper emphasized that physicians needed to know chemistry, and all chemistry regarding plants and people, in order to succeed at medicine. But he reminded his listeners at the conclusion of his speech that those physicians should also not make chemistry more complicated than it needed to be in order to help the patient. He used the work of Marcet to make the point that chemistry was portable and not a secretive enterprise [103]:

> I agree with that most able physician and chemist, Dr. Marcet (2 Med. and Chir. 'Trans, p. 358) that the large and dismal subterraneous laboratory of the old chemists, is now changed for the fire side of a comfortable study; and that under the auspices of Dr. Wollaston and two or three more of the British chemists, the analysis of small quantities of matter with neatness and accuracy, promises to give an essential impulse to the progress

> of analytical chemistry. In fact, the apparatus for experiments in medical chemistry ought to occupy no more space than the drawer of a book case, and the required investigations may be prosecuted without injury to a mahogany table by the fire side.

He was also interested in the collection building that both Marcet and Wollaston had engaged in regarding urinary calculi. But Cooper turned most of his chemical work to practical guides to chemistry, chemical analysis, and engaging in the rhetoric to promote it. And he summed up and punctuated the importance of chemistry to medicine as [104]: "INDISPENSIBLE to medicine." He thought that physicians coming after him would not see this dichotomy.

5.5 Cooper, Malaria, and Philadelphia

Cooper also wanted to explain contagion through chemistry and humoral medicine. He discussed his ideas about malaria in his works on politics and political economy. He theorized that people and nature were intertwined and part of the human experience. Cooper grieved that [105], "It is impossible to free the condition of man from disease and death," and he added that, "If nature hath ordained the laws of population to overrun, when it is unchecked, the law of production and subsistence, she must have provided also the counter balance." He wanted to understand nature and chemistry to ameliorate the human condition that seemed destined to suffer from diseases. Cooper also speculated that changes in city living could end suffering, and he proposed guidelines that could end diseases in cities. He cited malaria as a test case to illustrate his proposed guidelines. After giving a short history of Malaria as a disease that originated in Europe, Cooper discussed its presence in the swamplands of the South. He called for swamps to be drained and anything that was decomposing in those areas to be removed, as he thought that diseases originated from swamps and decomposing substances. From the decomposing of living substances like plants came the beginnings of other diseases like bilious fevers, especially in the hot months in the South.

But physicians were often not heard by townspeople, according to Cooper, and he drew an analogy to classical literature, with the story of Cassandra, the prophetess whose warnings were not heard [106], "people in general know so little of physiology and pathology, which the admonitions of the physician are like the prophecies of Cassandra, always unheeded." People would accidently release the decaying matter in swamps through cutting down trees, which released the miasmas into the airs, which sickened people by the miasmas affecting their internal fluids.

In his discussion of miasmas, Cooper turned to Philadelphia. Philadelphia was a city that had too much stagnate water, especially in regards to mud. Cooper called for the water to be removed and all stagnate water sources like swamps and other areas that were not free flowing to be eliminated. This was crucial to the health of the people of Philadelphia [106].

Other diseases that affected Philadelphia included yellow fever, and Cooper believed that a similar strategy could end its spread in the city. Cooper thought that the disease could be controlled by keeping the miasmas from spreading. Cooper invoked the ideas of Samuel Jackson, a physician who treated yellow fever in Philadelphia [107, 108]: "...who has shown in what way the yellow fever can be imprisoned and circumscribed until it be eradicated." Cooper promoted the suppression ideas of Samuel Jackson, such as building fences to keep miasmas from spreading and keeping suffers in areas with fresh air. Cooper proposed cleanliness and fresh air to keep miasmas from spreading and yellow fever was [109], "...contagious only when the patient is in a confined apartment, and the air is unchanged; but not contagious or infectious in a well aired apartment, with scrupulous attention to cleanliness." Cooper thought that people had some type of affect over the spread of miasmas.

In explaining miasmas, Cooper also had a humoral theory regarding the spread of disease. It was difficult for humoral theory to explain contagion, and the theory itself struggled to address contagious diseases throughout the nineteenth century [110, 111]. He speculated that miasmas might be living organisms. But Cooper also saw the affect of miasmas affecting poor people more than other groups [109], "The crowded and dirty state of jails, hospitals, poor houses, and boarding houses, is a proper subject of municipal control, and would certainly fall within the power and jurisdiction of a board of health." If the miasmas were not contained, the poor would make the rich sick [109], "The health of the rich is jeopardized by the filth of the poor." He also called for sick people to be isolated from others in Philadelphia.

Cooper further speculated about the nature of miasmata. In a footnote away from the main text, Cooper further thought of the miasmas as living creatures [112]:

> As the miasmata and effluvia of infectious diseases must come in contact with a body infected by them, generally by means of the Schneiderian membrane, they are, in a certain sense contagious. By contagious disorders, strictly speaking, I mean, *animal poisons communicable by contact, independent of breathing*. The more I consider this subject, the more I am persuaded that the whole tribe of infectious and contagious disorders owe their origin to animalculae; and that they have their infancy, their maturity, and their decline.

Throughout the text, Cooper continued to advocate that these living miasmata caused sickness. Cooper further emphasized his ideas to the reader [112]:

> The whole doctrine of equivocal generation requires to be reconsidered. Yellow fever first attacks the stomach, bilious fever the liver; black vomit examined by a microscope, presents a congeries of animalculae; the bubo of the plague is full of them; so are the pustules of psora. The rot in sheep seems to be owing to animalculae. If this opinion be well founded, no wonder that a chemical examination of air, cannot detect miasma which does not depend on the chemical state of the atmosphere.

Cooper's ideas framed the causes of diseases by issues with living miasmas, changes in the humors of the body, and rotting matter in the natural world.

5.6 The End of Cooper's Life

Toward the end of Thomas Cooper's life, he experienced many tribulations. While in South Carolina, Cooper changed from the anti-slavery advocate he had been earlier in his life to a supporter of the master class and slavery in South Carolina, winning favor with slave holders in the state. But he also continued to alienate religious people and clergy with his controversial ideas about the oldness of the earth [113]. Ultimately, Cooper had to quit his job as President at South Carolina College in 1833 because of many of his religious and philosophical ideas. Cooper continued to build support from the master class, supporting slavery, and also publishing edited books of documents pertaining to the laws of South Carolina [114, 115]. However, Cooper's health began to fail while editing the *Statues at Large*, a book that recorded all the legal decisions and law history for the state of South Carolina. Cooper started having problems with his heart and respiration, as well as with his eyesight. He corresponded with another South Carolina physician, Dr. M. H. DeLeon, who practiced medicine in Columbia. Cooper wrote his doctor and reported his feelings about his health [116]:

> You and I are by this time both sensible that my leaky etc. vessel has received a shot between wind and water, that leaves no alternative but whether the final submersion shall be effected by asthma or dropsy. As I have neither art nor part in this matter, I leave my two enemies to squabble in their own way. But I think I may venture to appeal to you, that however debilitated are my legs in the basement story, my brains in the attic have not received much injury.

Cooper died in 1839. His life was a project of linking together many projects of medicine and chemistry. He tried to weave together humoral ideas, chemical understandings of the body's fluids, and explanations of how the natural world created contagion and diseases like yellow fever and miasmas [117, 118].

Cooper was an emotional defender of his scientific, medical, chemical, and political ideas. Interestingly, many of these ideas would shift to fit his immediate needs. Cooper wanted to revive humoral theory and saw its high value in explaining contagion and diseases, as well as how the natural world invokes diseases in the fluids of the body. The chemically commitment physician could also adjust the fluids of the body to remove them from a diseased state. Cooper was also well read and engaged with the latest medico-chemical literature from Europe, such as the work of Thomas Thomson and Alexander Marcet. Something uniquely American was at play with the work of both Thomas Cooper and Edward Darrell Smith in their strong and clear calls for the return of humoral pathology. European chemists were interested in humoral pathology, and they did not make as grand declaration as Cooper and some of his American peers.

European chemists, physicians, and surgeons were interested in understanding the chemistry of the body, especially how its fluids responded to disease. Physicians and surgeons like Alexander Marcet and Astley Cooper wanted to chemically understanding the blockages that appeared in the body, such as urinary stones, in order to improve lithotomies and hoped to prevent their occurrences all together. The next chapter shifts to Europe and medico-chemical analysis of urinary stones, which could be seen as the application of many of the optimistic ideas of Thomas Cooper in the relationship between chemistry and medicine.

References

1. Cooper T (1818) A discourse of the connexion between chemistry and medicine. Abraham Small, Philadelphia, p 18
2. Malone D (1926) The public life of Thomas Cooper (1783–1839). Yale University Press, New Haven
3. Freehling WW (1991) The road to disunion Vol 1: Secessionists at Bay, 1776–1854. Oxford University Press, New York
4. Freehling WW (1965) Prelude to Civil War: the nullification controversy in South Carolina, 1816–1836. Oxford University Press, New York
5. Malone D (1926) A fighting modernist. VA Qrt Rev 2:49–62
6. Freehling WW (1991) The road to disunion Vol 1: Secessionists at Bay, 1776–1854. Oxford University Press, New York, p 256
7. Lander EM (1953) Dr. Thomas Cooper's views in retirement. SC Hist Mag 54:173–184
8. Kilbride D (1993) Slavery and utilitarianism: Thomas Cooper and the mind of the Old South. J South Hist 59:469–486
9. Schofield RE (1997) The enlightened Joseph Priestly: a study of his life and work from 1773 to 1804. Penn State Press, University Park
10. Cohen SS (2003) The correspondence of Thomas Jefferson and Thomas Cooper. Proc Am Philos Soc 147:39–64
11. Dreisbach DL (ed) (1996) Religion and politics in the Early Republic: and the Church State Debate. University Press of Kentucky, Lexington, p 29, note 27
12. Himes CF (1918) Life and times of Judge Thomas Cooper: jurist, scientist, educator, author, publicist. Dickinson School of Law, Carlisle
13. Rosenberg C (1962/1987) The cholera years: the United States in 1832, 1849, and 1866. University of Chicago Press, Chicago, p 6
14. Rosenberg C (1977) The therapeutic revolution: medicine, meaning, and social change in nineteenth century America. Persp Bio Med 20:485–506
15. Johnson N (1991) Nineteenth-century rhetoric in North America. Southern Illinois University Press, Carbondale
16. Pender S, Struever NS (2012) Rhetoric and medicine in early modern Europe. Ashgate Publishing Limited, Burlington
17. Malone D (1948) Jefferson and his time. University of Virginia Press, Charlottesville
18. Ellis HM (1920) Thomas Cooper—a survey of his life. S Atlantic Qrtly 19:24–42
19. Prichard RW (1999) History of the Episcopal Church, Revised edn. Morehouse Publishing, Harrisburg
20. Dorn C (2017) For the common good: a new history of higher education in America. Cornell University Press, Ithaca
21. Lehman FK (2008) "Seditious Libel" on Trial, Political Dissent on the Record: "An Account of the Trial of Thomas Cooper" as Campaign Literature. Penn Mg of Hist and Bio 132:117–139

22. Ellis HM (1920) Thomas Cooper—a survey of his life. S Atlantic Qrtly 19:25
23. Ellis HM (1920) Thomas Cooper—a survey of his life. S Atlantic Qrtly 19:26
24. Cooper T (1815) A practical treatise on dyeing, and callicoe printing. Thomas Dobson, Philadelphia
25. Armstrong EV (1937) Thomas Cooper as an iterant chemist. J Chem Educ 14:153–158
26. Edelstein SM (1948) The role of chemistry in the development of dyeing and bleaching. J Chem Educ 25:144–149
27. Ellis HM (1920) Thomas Cooper—a survey of his life. S Atlantic Qrtly 19:28–29
28. Cooper T (1787) Letters on the slave trade. C. Wheeler, Manchester
29. Swaminathan S (2009) Debating the slave trade: rhetoric of British national identity, 1759–1815. Routledge, London
30. Cohen SS (1982) Two refugee chemists in the United States, 1794: how we see them. Proc Am Philos Soc 126:301–315
31. Cooper T (1794) Some information respecting America. J Johnson, London
32. Cooper T (ed) (1807) The memoirs of Joseph Priestly. E. Hemsted, London
33. Cohen SS (2003) The Correspondence of Thomas Jefferson and Thomas Cooper: a previously unpublished manuscript of Dumas Malone. Proc Am Philos Soc 147:39–64
34. Cooper T (1800) An account of the trial of Thomas Cooper, of Northumberland: on a charge of libel against the president of the United States. J. Bioren, Philadelphia
35. Hoffer PC (2011) The Free Press crisis of 1800: Thomas Cooper's trial for seditious libel
36. Bird W (2020) Criminal dissent: prosecutions under the Alien and Sedition Acts of 1798. Harvard University Press, Cambridge
37. Bell WJ Jr (1953) Thomas cooper as professor of chemistry at Dickinson College 1811–1815. J Hist Med Allied Sci 8:70–87
38. Marsden GM (2021) The Soul of the American University revisited: from Protestant to postsecular. Oxford University Press, Oxford
39. Ragosta JA, Onuf PS, O'Shaughnessy AJ (ed) (2019) The founding of Thomas Jefferson's University. University of Virginia Press
40. Smith EF (1917) Life of Robert Hare, 1781–1858. J. B. Lippincott Company, Philadelphia
41. Grasso C (2018) Skepticism and American Faith: from The Revolution to the Civil War. Oxford University Press, Oxford
42. LaBrode M (1874) History of South Carolina College: from its incorporation, Dec. 19, 1801 to Dec. 19, 1865. Walker Evans & Cogswell Printers, Charleston
43. Hungerpiller JC (1917) A sketch of the life and character of Johnathan Maxcy, D. D. M. A. thesis, University of South Carolina, Columbia
44. Cooper T (1818) A discourse of the connexion between chemistry and medicine. Abraham Small, Philadelphia, p 19
45. Frank JP (1817) System Einer Vollstandigen Medicinishen Polizey. Carl Ghaumburg and Comp, Wien, p 42
46. Driggers EA (2021) "The voice of the stomach": the mind, hypochdriasis and theories of dyspepsia in the nineteenth century. Hist of Pysc 32:58–99
47. Cooper T (1818) A discourse of the connexion between chemistry and medicine. Abraham Small, Philadelphia, p 20
48. Rosenberg CE (1989) Body and mind in nineteenth-century medicine: some clinical origins of the neurosis construct. Bull Hist Med 63:185–197
49. Dyde S (2015) Cullen, a cautionary tale. Med Hist 59:222–240
50. Cooper T (1818) A discourse of the connexion between chemistry and medicine. Abraham Small, Philadelphia, p 21
51. Lawrence SC (1996) Charitable knowledge: hospital pupils and practitioners in eighteenth century London. Cambridge University Press, Cambridge
52. Cooper T (1818) A discourse of the connexion between chemistry and medicine. Abraham Small, Philadelphia, pp 22–24
53. Arikha N (2007) Passions and tempers: a history of the humors. Ecco, New York

54. Porter R (1997) The greatest benefit to mankind: a medical history of humanity. W. W. Norton, New York
55. Ladenburg A (1908) lectures on the history of the development of chemistry since the time of Lavoisier trans. Dobbin L. University of Chicago Press, Chicago, p 87
56. Cooper T (1818) A discourse of the connexion between chemistry and medicine. Abraham Small, Philadelphia, p 24
57. Chang K (2011) Alchemy as studies of life and matter: reconsidering the place of vitalism in early modern chemistry. Isis 102:322–329
58. Brown TM (1987) Medicine in the shadow of the Principia. J Hist Idea 48:629–648
59. Dear P (2006) The ineligibility of nature: how science makes sense of the world. University of Chicago Press, Chicago
60. Kim MG (2003) Affinity, that Elusive Dream: a genealogy of the chemical revolution. MIT Press, Cambridge
61. Levere TH (2001) Transforming matter: a history of chemistry from alchemy to the buckyball. The Johns Hopkins Press, Baltimore
62. Cooper T (1818) A discourse of the connexion between chemistry and medicine. Abraham Small, Philadelphia, p 25
63. Novikoff AJ (2013) The medieval culture of disputation: pedagogy, practice, and performance. University of Pennsylvania, Philadelphia
64. Lim R (1995) Public disputation, power, and social order in late antiquity. University of California Press, Berkley
65. Rodda J (2016) Public religious disputation in England, 1558–1626. Routledge, New York
66. Young JR (1999) Domesticating slavery: the master class in Georgia and South Carolina, 1670–1837. University of North Carolina Press, Chapel Hill
67. West JB (2014) Joseph Priestley, oxygen, and the enlightenment. Am J Physiol Lung Cell Mol Physiol 306:111–119
68. Johnson N (1991) Nineteenth-century rhetoric in North America. Southern Illinois University Press, Carbondale, p 31
69. Cooper T (1818) A discourse of the connexion between chemistry and medicine. Abraham Small, Philadelphia, p 26
70. Johnson S (2008) The invention of air: a story of science, faith, revolution, and the birth of America. Riverhead Books, New York
71. Starr D (1998) Blood: an epic history of medicine and commerce. Knopf Doubleday Publishing, New York
72. Cooper T (1818) A discourse of the connexion between chemistry and medicine. Abraham Small, Philadelphia, p 28
73. Scull A (2009) Hysteria: the disturbing history. Oxford University Press, New York
74. Bono JJ (1990) Reform and the languages of renaissance theoretical medicine: Harvey versus Fernel. J Hist Bio 23:341–387
75. Cooper T (1818) A discourse of the connexion between chemistry and medicine. Abraham Small, Philadelphia, p 37
76. Broussais FJV (1833) On irritation and insanity, ed. Cooper T. R. Hunter, London
77. Proman PE (ed) (2018) The Cambridge Companion to Hippocrates. Cambridge University Press, New York
78. Dawson I (2005) Medicine in the middle ages. Enchanted Lion Books, New York
79. Pormann PE, Savage-Smith E (eds) (2007) Medieval Islamic medicine. Georgetown University Press, Washington, DC
80. Nutton V (2004) Ancient medicine. Routledge, New York
81. Fox RL (2020) The invention of medicine: from Homer to Hippocrates. Basic Books, New York
82. Chadwick J, Mann WN trans (1978) Hippocratic writings. Penguin Books, New York, p 148
83. Langholf V (1990) Medical theories in Hippocrates: early texts and the "Epidemics." Walter de Gruyter, Berlin
84. Tannenbaum JR (2012) Health and wellness in colonial America. Greenwood, New York

85. Golkinski J (2007) British weather and the climate of enlightenment. University of Chicago Press, Chicago
86. Johnson S (2006) The ghost map: the story of London's most terrifying epidemic: and how it changed science, cities, and the modern world. Penguin, New York
87. Driggers EA (2022) "…[T]he movement of a celestial system than a human invention:" Abram Blanding and bringing water to Columbia. Water Hist 14:21–40
88. Nelson MK (2002) The landscape of disease: swamps and medical discourse in the American Southeast, 1800–1880. Miss Qrtly 55:535–567
89. Bynum (1981) Cullen and the study of fevers in Britain, 1760–1820. Med Hist Sup 1:135–147
90. Willoughby UE (2017) Yellow fever, race, and ecology in nineteenth-century New Orleans. Louisiana State University Press, Baton Rouge
91. Cooper T (1818) A discourse of the connexion between chemistry and medicine. Abraham Small, Philadelphia, p 34
92. Heyne B (1814) Tracts, historical and statistical, on India. Robert Baldwin, London, pp 181–185
93. Hurst JW, Fye WB (1997) Jean Franclois Fernel. Clin Cardiol 20:1037–1038
94. Allbutt TC (1900) Medicine in 1800. Brit Med J 2:1848–1852
95. Polianski IJ (2021) Airborne infection with Covid-19? A historical look at a current controversy. Microbes Infect 29:10485
96. Cooper T (1818) A discourse of the connexion between chemistry and medicine. Abraham Small, Philadelphia, p 38
97. Chapman N (1817) Discourses on the elements of therapeutics and Materia medica, vol 1. James Webster, Philadelphia
98. Chapman N (1817) Discourses on the elements of therapeutics and Materia medica, vol 1. James Webster, Philadelphia, p 25
99. Richman I (1967) The brightest ornament: a biography of Nathaniel Chapman, M.D. Pennsylvania Heritage, Inc., Bellefonte. Citation found in Bibliography of the history of Medicine 1964–1969 by U.S. Department of Health, Education, and Welfare
100. Richman I (1966) Nathaniel Chapman, M.D. Penn Med 69:58–9. Citation found in Bibliography of the history of Medicine 1964–1969 by U.S. Department of Health, Education, and Welfare
101. French R, Wear A (eds) (1991) British medicine in an age of reform. Routledge, New York
102. Cooper T (1818) A discourse of the connexion between chemistry and medicine. Abraham Small, Philadelphia, p 39
103. Cooper T (1818) A discourse of the connexion between chemistry and medicine. Abraham Small, Philadelphia, pp 47–48
104. Cooper T (1818) A discourse of the connexion between chemistry and medicine. Abraham Small, Philadelphia, p 48
105. Cooper T (1831) Lectures on the elements of political economy, 2nd edn. M^C Morris and Wilson, London, pp 335–336
106. Cooper T (1831) Lectures on the elements of political economy, 2nd edn. M^C Morris and Wilson, London, p 338
107. Cooper T (1831) Lectures on the elements of political economy, 2nd Edn. M^C Morris and Wilson, London, pp 338–339
108. Jackson S (1821) An Account of the Yellow or Malignant Fever, As It Occurred in the City of Philadelphia in 1820. M. Carey & Sons, Philadelphia
109. Cooper T (1831) Lectures on the elements of political economy, 2nd Edn. M^C Morris and Wilson, London, p 339
110. Santer M (2015) Confronting contagion: our evolving understanding of disease. Oxford University Press, New York
111. Rosenberg CE (1962) The Cholera years: the United States in 1832, 1849, and 1866. University of Chicago Press, Chicago
112. Cooper T (1831) Lectures on the elements of political economy, 2nd edn. M^C Morris and Wilson, London, pp 339–340, footnote

113. Wright C (1941) The religion of geology. New Eng Qrtly 14:335–358
114. Cooper T (1824) A tract on the proposed alteration of the tariff. Joseph R. A. Skerret, Philadelphia
115. Fox-Genovese E, Genovese ED (2005) The mind of the master class: history and faith in the Southern Slaveholder's Worldview. Cambridge University Press, New York
116. Waring IJ (1964) A history of medicine in South Carolina:1670–1825. South Carolina Medical Association, Charleston, p 202
117. Lippincott G (2003) Rhetorical chemistry: negotiating gendered audience in nineteenth-century nutrition studies. J Bus Tech Comm 17:10–49
118. Richard J, Morrman H (1967) A history of the Church in England. University of California, Berkley

Partnerships Between Surgeons and Chemists

6

Alexander Marcet (Fig. 6.1) was a political dissident, chemist, and physician, who operated in intellectual circles in the nineteenth century. Originally from Geneva, he emigrated to England, where he practiced both medicine and chemistry [1, 2]. Marcet turned his curiosity to compiling a comprehensive study of the history of urinary stones and their chemical character. His 1817 study *An Essay on the Chemical History and Medical Treatment of Calculous Disorders* was popular enough to be transcribed into German and French[1] [3]. Marcet wanted to do one of the first comprehensive studies of the chemistry of urinary stones in a similar fashion to William Hyde Wollaston's article. Wollaston was a physician and chemist who discussed ideas regarding the chemical classifications of stones. His work was of interest to chemically informed physicians like Edward Darrell Smith and Thomas Cooper, but also surgeons who wished to save their patients from the painful, and dangerous, operation for the stone, called the lithotomy. Marcet, because of his expansive social circles, struck up a friendship with the celebrity surgeon Sir Astley Cooper.

The line between surgery and medicine was often set on the operation for the stone. In the original oath attributed to Hippocrates, it forbids physicians from performing lithotomy. The ancient oath included the line [4, 5]: "I will not use the knife, even upon those suffering from stones, but I will leave this to those who are trained in this craft." The modern oath often does not include any direct cautions about lithotomies [6]. Surgeons wanted to improve their hallmark surgery, as surgeons knew that lithotomies were in demand from patients, despite the potential pain [7–10]. Surgeons have been collecting the histories of the lithotomy since the early nineteenth century [10].

[1] These editions are entitled *Chemische Untersuchungen über die Harnsteine* and *Essai sur l'Histoire Chimique des Calculs et sur le Traitement médical des Affections Calculeuses.*

E. A. Driggers, *Early Nineteenth Century Chemistry and the Analysis of Urinary Stones*, Perspectives on the History of Chemistry,
https://doi.org/10.1007/978-3-031-34973-7_6

Fig. 6.1 Alexander John Gaspard Marcet. Stipple engraving by H. Meyer after Sir H. Raeburn (Wellcome Collection)

The partnership between Marcet and Cooper led to a more detailed understanding of what type of stones patients were suffering from, the statistics related to the mortality of the lithotomy, and Cooper also provided Marcet with a collection of removed urinary stones that he could test with his chemical tools to understand the history of the stones that the patients in Norfolk, England suffered from. Both medico-chemists like Marcet and Astley Cooper also saw the body in humoral terms, focusing on the fluids of the bodies as the cause for disease.

Building collections of stones were crucial to furthering the work of chemists to understand the stones of the body, and surgeons were important in providing those items. Social scientists like Susan Leigh Star and James R. Griesmer would see the historical discussions of urinary stones by surgeons, chemists, and physicians as being centered on an object that was of interest to all groups and required them to go beyond their professional boundaries. Bodily stones required medico-chemists and surgeons to [11]: "…translate, negotiate, debate, triangulate, and simplify in order to work together." Collaborations, such as those with surgery, medicine, and chemistry, move ideas and information around these communities and creates new ways of knowing, such as the increase in the chemical histories of urinary stones that medico-chemists like Alexander Marcet was able to produce. It was likely that surgeons like Cooper could use the chemical knowledge and potential interventions to improve his surgical practice.

Surgery and medicine are often partitioned in healthcare history, as they were different: medicine was a profession, and surgery was a craft or occupation; but they are both professions now. Aspiring surgeons apprenticed themselves to established surgeons, while doctors often went to universities, though this was not the case with all physicians [12]. Chemistry was often approached timidly by surgeons as recently as 1965, with articles and other publications skeptically cautioning surgeons of the [13] "Use and Abuse of Clinical Chemistry in Surgery." Historians of medicine are still adding the contributions of surgeons to the story of health care in the nineteenth century. While other historians do not think that the divide between medicine and surgery was as great as current historians believed it was in the nineteenth century [14], William Bynum and Roy Porter warned historians of medicine that [15], "several contributors argue that historians often hamper themselves with anachronistic or inappropriate models of medicine, implicitly borrowed from the nineteenth century." Their quote has the warning that readers today buy into the division that physicians wanted to make as they tried to make their argument for professionalized medicine [16, 17]. Historians, though not enough, have been repositioning the role of surgeons as contributing to scientific medicine over the last thirty years, especially in their role in participating in the European Enlightenment [18]. But more works on the scientific contributions of surgeons needs to be added to the historiography of both the European Enlightenment and the growth of scientific medicine during the nineteenth century; many scholars of surgery today call for their current contributions and renewed interest in science [19]. Scholars like Phillip Wilson have astutely found that many surgeons contributed to the pages of the journals of top intellectual societies of the eighteenth and nineteenth century, such as the Royal Society's *Philosophical Transactions* [18]. But Wilson also adds that those contributions started to trend down. But in this book, surgeons are highly involved with studying and analyzing urinary stones, as well as other stones produced in the body. Astley Cooper is an excellent example, and his partnership with Marcet shows the ineffectiveness of the dichotomy of medico-chemistry and surgery. Studying their partnership reveals how the chemical analysis of stones worked as two fields concentrated on shared questions: how do stones form in the body, and is there anything anyone can do about them?

6.1 The Character of Alexander Marcet

Marcet did not start out his career destined to become a physician and chemist when he started his life in Switzerland in 1770 [1, 2, 20, 21].[2] He was born to protestants in Switzerland, Huguenots, who had been involved in business. His

[2] Marcet is also listed as a biographical entry in the *Oxford Dictionary of National Biography*. There is, unfortunately, no book length study of all his many contributions to science, nor Jane Marcet.

family was mostly merchants. Alexander Gispard Marcet broke with family tradition and with their blessing, started studying science. However, his studies were interrupted when conscription occurred in Geneva because of the French Revolution, and it appeared that Marcet was going to have to serve in the military. The French Revolution became more radicalized and Marcet decided to leave on his own free will or was ultimately forced out of Geneva during 1794. Scholars continue to question whether he was removed and but was not re-admitted to his former nation later.[3] The French government forcibly incorporated Geneva into its new empire [22–25]. Marcet was not able to return to his home nation for around five years, and his fellow soldiers and friends from his early life were also forbidden to return. Marcet, after his travel ban, decided to study at the University of Edinburgh, where he received his doctorate in medicine in 1797, where he produced his inaugural thesis on the subject of diabetes. Biographers during the time of his death wrote about the importance of his dissertation contributing to chemical knowledge [26], "He manifested his predilection for chemistry by selecting Diabetes as the subject of his inaugural dissertation; a disease concerning which some very plausible chemical theories had been recently proposed, and at that time engaged considerable attention at the university." After receiving his doctorate in medicine, Marcet landed in London, where he got a job at a public dispensary serving the needs of people on Carey Street as a doctor. He formally aligned himself with the Royal College of Physicians, receiving their "Licenate" in 1799. In that same year, he married the former Jane Haldimand. She came from a very wealth background, like Marcet, of merchants from Switzerland. The marriage produced three children, including Francois Marcet, who also chose to study medicine. Marcet formally became a citizen of the United Kingdom in 1800 and was working as a doctor at Guy's Hospital in London in 1804. He continued his interest and participation in science, and he joined the Royal Society in 1808 as a fellow of that institution.

Jane Haldimand Marcet was an incredibly interesting character in her own right in the nineteenth century. Jane Marcet contributed to scientific literature with her *Conversations*, which were several books that taught major subjects like chemistry, botany, and political economy to children and women [27, 28]. Aspiring natural philosophers, such as Michael Faraday, studied Marcet's work to learn chemistry and become acquainted with the field [29].

As wars continued in France with the rise of Napoleon and the quest to build a trans-Atlantic Empire, Marcet served the British in the Wars of Coalition. He served as an army physician in Portsmouth. During his wartime service, Marcet was stricken with fever, but ultimately beat the disease. Many of Marcet's colleagues praised his military service [30]:

[3] From the biographical scholarship it seems that Marcet was on the wrong side of the French Revolution in Geneva, it appears that he was against the radicalism of the Revolution and for Genevan independence.

> At the time when the Walcheren fever was committing dreadful ravages among our troops on their return from the expedition to Holland, in 1809, the want of additional medical assistance being urgently felt, Dr. Marcet volunteered his services, and was appointed to the superintendence of the General Military Hospital at Portsmouth; a duty which he performed with unremitting zeal, and which was interrupted only by himself becoming the subject of a similar disease. He was very severely affected, and recovered from it with great difficulty.

While serving the military, Marcet also taught students at Guy's Hospital in London. He worked there from 1807–1820, where he taught chemistry classes. Marcet's chemical teaching received a lot of positive feedback, and he built relationships with other chemists like J. J. Berzelius.

It was while serving as a teacher at Guy's Hospital that Marcet worked on his major research project on urinary stones, which was the culmination of his chemical tests and partnership with Astley Cooper. *An Essay on the Chemical History and Medical Treatment of Calculous Disorders* was published in 1817 after which other translations of the book appeared in Germany, France, and Sweden. Marcet passed away in 1822 from a disease that attacked his intestines and stomach. Jane Marcet published chemical literature after Marcet's death, with her famous *Conversations on Chemistry* published during 1832, and she then passed away in 1858.

6.2 Marcet's Chemical Research into Urinary Stones

An Essay on the Chemical History and Medical Treatment of Calculous Disorders is a real snapshot or freeze frame into the world of the chemical studies of urinary stones. The book also brings a wealth of case studies and experiential information to both chemists, surgeons, and physicians studying urinary stones. The question of how to improve knowledge regarding urinary stones brought chemists, surgeons, and physicians together in a communal goal. Marcet also refers to a lot of publications from all three disciplines. This fit Marcet's experiences as he held close friendships with members of all groups.

The inspiration for Marcet's work was William Hyde Wollaston, and Marcet praised Wollaston's medico-chemical work, especially his work regarding the stones of the body, including urinary stones. Marcet also shared a friendship with Wollaston, as they both were members and active in the Medico-Chirurgical Society [31]. The Medico-Chirurgical Society became an active organization that was interested in medico-chemistry and published papers from many surgeons, chemists, and physicians, and this is examined in the next to last chapter in the book.

Marcet frames his book not only as an intellectual study aimed at the "medical profession" but also as a teaching resource. In the plates of the book, he lists out the tools that the physician can use at the patient's bedside to examine the patient's urine and stones that are expelled. He outlined the tools chemists, physicians, and surgeons needed to perform chemical analysis. Those tools included a set of tongs, a heat source (such as a lamp), a blowpipe, as well as bottles for chemical reagents and analysis, a laboratory bench, and other glasses bottles to collect samples and

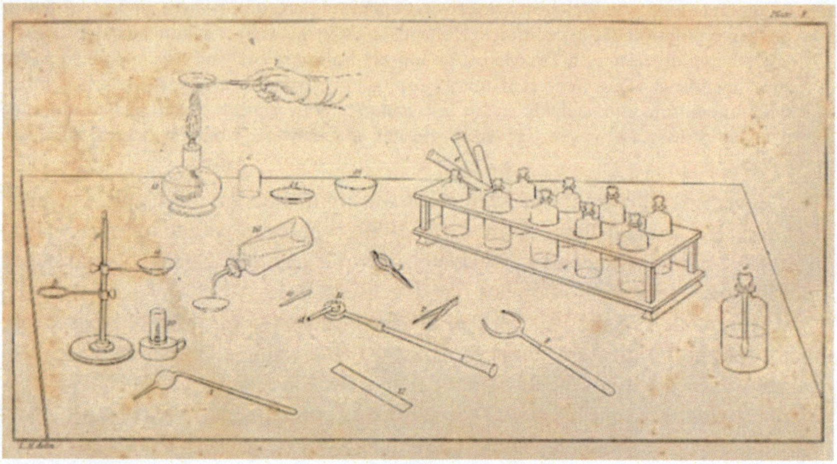

Fig. 6.2 Physician's chemical bench (Science History Institute)

hold other chemicals (Fig. 6.2). But Marcet did not want the student, or the professional wanting to expand their analytical abilities, to be restricted by needing a lot of equipment. Marcet felt that anyone could easily gather these tools to engaging in chemical studies of stones. *An Essay on the Chemical History and Medical Treatment of Calculous Disorders* contained pictures of all the needed tools for the physician to chemically examine their patient's expelled urinary stones.

Though Marcet was excited about the potential that chemistry could have in improving the treatment of bodily stones, he was cautious in his optimism: chemistry was not a cure, but could only provide comfort from pain. He was not advocating that chemistry could either end stones or be a substitute for the proven therapy of lithotomy. The value of chemistry to treating stones was high and perhaps momentarily unknown [32], "But if the progress of the disease may be arrested in its earlier stages, and if the pain and danger of a formidable operation can be averted*; or if after the operation, the tendency to a relapse may be effectually checked, enough, no doubt, will be gained to entitle the subject to our most serious attention." Chemistry was conceivably a tool that could stop internal processes from producing stones or continually chronic cases of stones.

The book was also a teaching manual to get medical students and interested physicians interested in learning about the tools of chemistry. He thought that his book would drive excitement about chemical research [33]:

> The practical utility of the pursuit, and the great facility with which these bodies may now be analyzed and discriminated from each other even by persons unaccustomed to chemical manipulations and the remarkable simplicity which modern chemistry has introduced in the history of calculi, compared to the singular scantiness of information which prevailed in this respect twenty or thirty years ago....

Marcet was a champion of the collaboration between surgery and medicine, and they both had a lot to add to each other's practice [34], "This tract can scarcely be viewed, even by those who wish to preserve with the strictest rigour the artificial line by which the different departments of the profession are circumscribed, as an encroachment upon the practice of surgery." He praised the maxim that they should work across any professional lines for the benefit of the patient [34], "Indeed such is the unavoidable and constant dependence of the professions of medicine and surgery on each other, that any apology on the subject would, in my opinion, be a kind of insult to the sense, or candour of my medical brethren." In general, he was skeptical of professional boundaries in the pursuit of knowledge [34], "Whatever advantage may arise from dividing and circumscribing the labours of the two professions, in practice, the greatest benefit may be expected, in science, from combining the studies of both." Marcet even created his own surgical tools. And he believed that medicine and surgery made each other better [34]:

> a physician, if conversant with the surgeon, will direct the effects of medicine with greater certainty and success while a surgeon will derive incalculable advantages in the treatment of local diseases from the knowledge he may have acquired of pathological principles. And the latter will soon learn in studying the phenomena and treatment of diseases, the fallacy and danger of the popular notion that the knowledge of the structure of the body is sufficient to enable us to obviate the disease to which it is liable.

Both medicine and surgery needed to work together for the common good. Alexander Marcet was also very thankful to Cooper for providing him access to the stones that he had extracted surgically from patients so that he could perform his chemical experiments.

Marcet also believed that urinary stones affected men and women differently. He thought because of different in physiology and constitution, women could cease suffering from the stone because of the statistical information that Marcet collected. But he soft peddled this idea not in the main text of the book. However, Marcet's statistical work was very important as he found out the nature of the danger of lithotomies. He determined that for every five patients that went under the knife for the lithomy, one patient would pass away from the surgery. In terms of general admissions at the hospitals in Norfolk, as well as Norwich, Marcet found that around one-third of the admissions to the hospital were cases of urinary stones, which resulted to around four hundred admissions that Marcet collected. He found that women were the smallest amount of admissions for the stone and also had the smallest amount of surgeries for stones. Because of the women's physiology, specifically her urethra, it allowed for women to avoid surgery and often had manual removal. He remarked that in regards to women [35]:

> The remarkable degree to which the urethra in females is capable of being dilated by well known mechanical means, renders it practicable in every instance to extract from their bladder, without the assistance of the knife any calculus of moderate size; and even stones of a very considerable bulk in a few instances, have been extracted in that manner.

Marcet continued to make this shadow claim about women not suffering from the stone in the future because of the optimism of the chemical work of himself and other medico-chemists, like Thomas Cooper and John Yelloly, eventually producing a remedy to extinguish stones in the body, or make them so small that they are removed by the surgeon's hand without the need of surgery.

6.3 Marcet Discusses the Chemical Nature of the Stones of the Body

Marcet's *Essay* is a book length project that focuses on explaining urinary stones comprehensively; it is a higher stakes version of the Wollaston article that discussed the taxonomies of stones earlier in this book. The *Essay* is an encyclopedic study of urinary stones for the medico-chemical specialist, but it is also designed for the neophyte physician wanting to learn bedside chemical analysis of urinary stones in order to assist patients. In setting the focus of his book, Marcet gives a comprehensive definition of urinary stones [36]:

> the formation of concretions in the urinary passages being occasioned by the separation and consolidation of certain ingredients contained in the urine, and being independent of any specific agency of the urinary organs themselves calculi are liable to form in any of the cavities to which the urine has access.

Urinary stones had their beginnings in the kidneys and could also be found in the bladder, as well as appear in urethras and ureter passages. Marcet laid out the steps of urination, where urine moved into the pelvis from other parts of the body, like the kidneys, and then it was removed from the body via the ureters. The ureters, according to Marcet, acted as a filtration system of the body. It was by filtering the urine of the body that stones appeared in the ureters or moved to other parts of the body through the ureters. Stones in the body appeared in the kidneys and in other parts of the body. Marcet wanted a visual representation of stones in the body so that the student could learn. One plate of the essay shows the stones stuck in the kidney (Fig. 6.3), what Marcet calls a [37] "morbid affection." The "cavities" of the body enlarge themselves when they are filled with stones. Interestingly, Marcet includes the detail that though the patient under his care died of kidney disease, and his kidney is represented in the plates, the patient never complained of pain in the kidney area. In *An Essay on the Chemical History and Medical Treatment of Calculous Disorders*, Marcet was able to have these diseases vividly portrayed in the plates included in the back of the book.

Additional plates in the book show how stones appeared in other parts of the body (Figs. 6.4 and 6.5). Though urinary stones appeared in different parts of the body, Marcet believed that they began in the kidney and the patient's bladder. But when stones traveled to other areas of the body, they would reap havoc for the patient. Marcet wanted to show how organs and passages of the body could be expanded to accommodate surprisingly large stones. Not only does Marcet's

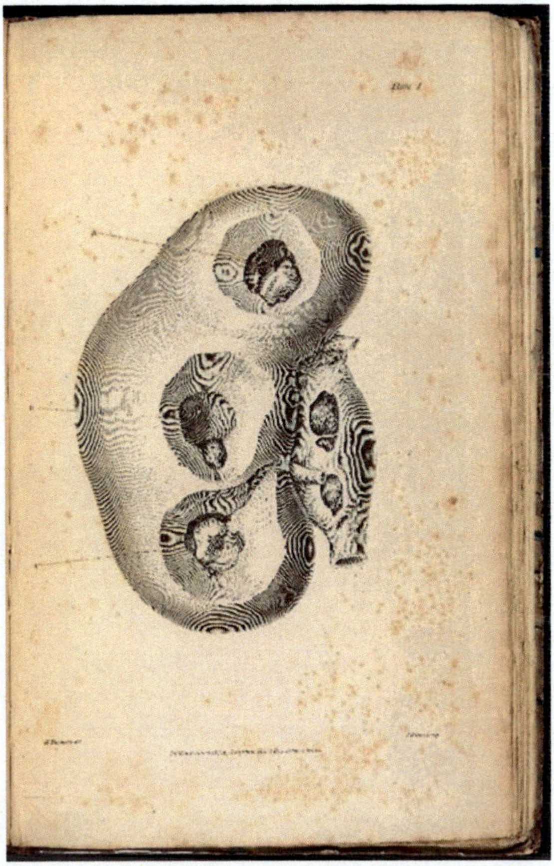

Fig. 6.3 Diseased kidney with stones (Science History Institute)

Essay include images to teach the physician and medico-chemists about how the stones appeared in the body, they included experimentational information about case histories of urinary stones that Marcet either witnessed or complied from other physicians and surgeons.

Marcet wanted to share case histories not only because they were good for teaching medical students, but they were also good at instructing chemists. One story that Marcet decided to share was a long-suffering man who had suffered from urinary stones for more than ten years, expelling blood, pebbles, and mucus like secretions, but could urinate, so Marcet and the other attendance to the case did not think that he was suffering from urinary stones [38]. Without suffering from pain in the penis, not allowing doctors to probe him for urinary stones, the patient died in large amounts of distress for unclear reasons. Marcet performed an autopsy and found a large mass that had been encased in the body, but allowed urine to flow through the patient normally [39].

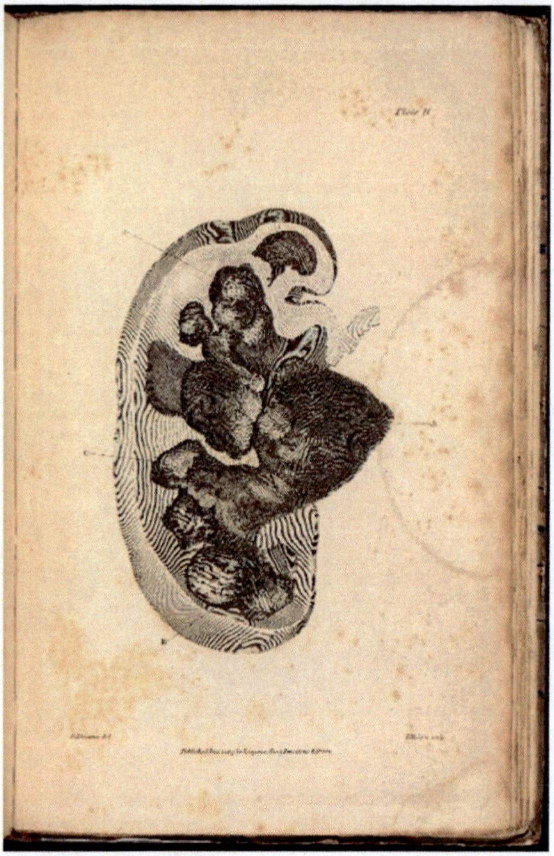

Fig. 6.4 Plate of a large stone from the collection of a surgeon (Science History Institute)

He included for the reader some characteristics to identify patients who might have urinary stones. There were a lot of characteristics that Marcet included, such as weakness, localized pain near the kidney, foul smelling urine, blood, and discolored urine produced by the patient. The patient suffered from pain when the stone traveled throughout the body, such as when stones left the kidneys, causing back pain or thigh pain as it traveled to the penis and then outside of the body. Patients also suffered the distresses of needing to urinate, but not being able to complete the action. A lot of patients suffered from pain in their ureters, but might not realize when the stone actually left the body, and would be only left with mucus or blood, or the feelings of inflamed body parts, or the patient could suffer from stones with unknown characteristics that often confused the doctor. Bodily irritation, like a lot of conditions in the nineteenth century, and irritation could be very painful [40]. Actions that patients perform daily could make their stone pain worse, like riding horses, and if irritation got worse in the body, like in the

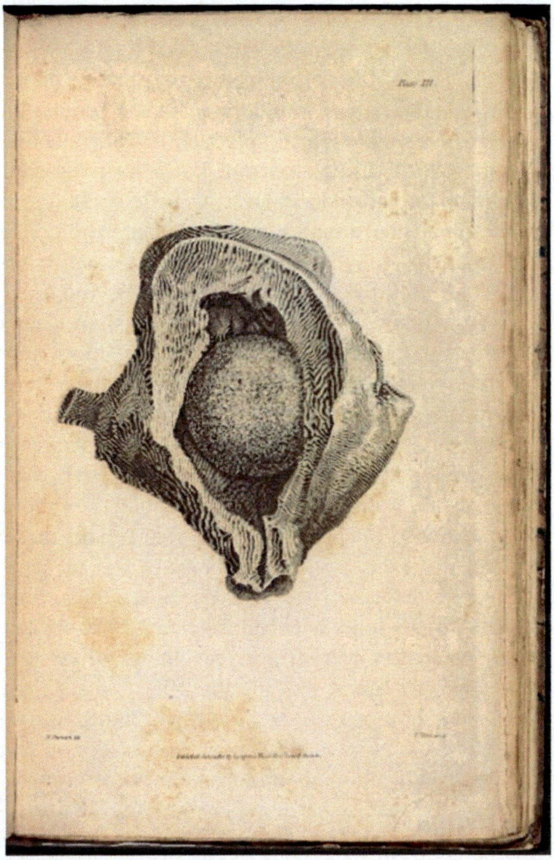

Fig. 6.5 Stone stuck in bladder, from Astley Cooper (Science History Institute)

bladder, it could spread to other parts of the body and kill organs, like the stomach. Stone could cause irritation in other parts of the body, like the urethra, and stones that were lodged in the prostate could hold up urination in the body as well. Stones were another characteristic of full body irritation that was thought to be a primary cause of disease. Marcet could be considered an adherent to some aspects of humoral pathology because of his ideas that the fluids of the body spread irritation through stones or irritations in the body produced stones. Marcet did not explicitly promote humoral pathology in the same manner as Thomas Cooper or Edward Darrell Smith in previous chapters in this book.

6.4 Questions About Stones

In his *Essay*, Marcet wanted to understand if nationality, or lifestyle, or bodily age, or other social habits affected the production of urinary stones. He wanted to bring a mathematical approach to understanding urinary stones. Marcet, living in the age of Enlightenment, put a lot of attention mathematizing the world in order to further understand it [41]. He wanted to understand mortality numbers in lithotomies, but found a lot of challenges in collecting information. At that point, no hospitals were really collecting information about who underwent lithotomies and how they fared after the surgery. He did find, however, that two hospitals did not fit that history, and these hospitals were located in Norfolk and Norwich, UK. Marcet was amazed that he could get information about surgeries performed at these hospitals going back about fifty years. Surgeons also had the many stones that were removed during those lithotomies.

Because of his social connections to surgeons through the military and his work in the Medico-Chirurgical Society, he had access to these surgeons, and they helped him compile information about the lithotomies performed at the Norfolk and Norwich hospitals. He was able to sort these details by gender and age. He found that men, not women, mostly had to have the surgery. And there were a lot of deaths associated with the lithotomies. Starting with Norfolk, that hospital had 18,859 admits between the years 1772–1816. Around 428 patients came into the hospital on average per year. In the 44 years of information that Marcet could examine, he found that around 506 of the 18,859 total admitted patients which came to the hospital for surgical lithotomies, which averaged to one patient in around 38 general admittances. He was able to present the numbers of lithotomies by year as well. At Norwich, there were around 100 admissions for lithotomies from 1772–1782, that number grew to 120 from 1782–1792, and those numbers slightly trended down from 1792–1802 to 116, but then increased again to 137 from 1802–1812 [42]. Marcet found that overall more lithotomies were growing, and he tried to compare this trend to other hospitals. He continued to ask for information from surgeons, who could only provide him with estimates, and he further probed information at Guy's Hospital in London, from which he received assistance from nuns working at the hospital. He constructed estimates for Guy's from his investigations. He determined that comparatively, Guy's Hospital in London had a patient go into the hospital for a lithotomy from around 300 admissions.

He also thought that weather had an effect on the production of urinary stones. But he also emphasized in his *Essay* that surgeons, physicians, and chemists should also collect information about lithotomies and urinary stones cases to build larger understandings of where patients were suffering from stones as he said that this was both good for humanity and [43], "advancement of science." Keeping a collection of stones from lithotomies was also a good idea because of the small amount of material needed to carry out chemical research on the nature of the stones removed. Marcet promoted the idea [43]:

> Any small fragment detached from one of these portions, or merely the sawings of the calculus, will in most every instance, be sufficient for chemical examination; and while the remaining half will afford a much more instructional preparation that the calculus in its entire state, the portion encroached upon will still furnish a useful duplicate.

He added that better chemistry would help patients as [43], "…the pathology and treatment of this [urinary stones] obscure disorder."

6.5 Assembling Knowledge About Stones Through Chemistry

Marcet proposed his own system of chemical classification of urinary stones. Previous systems of organization for urinary stones used the place of the body where the stone originated, which included the kidneys, urethra, or origination from the bladder. He did not feel that the older system was as accurate as a system based on chemistry. Marcet noted that organizing stones could end up in any part of the body. He proposed five types of urinary stones based on their chemical composition. The first group was lithic acid. The second group included those stones of phosphate of lime. The third group was the ammonia-magnesium. The fourth group consisted of stones consisting of the oxalate of lime. The final group was called the cystic oxyd. Marcet was open to additional classifications beyond these five groupings because stones could be made up of more than one substance.

In the first group, the lithic acid stones, there character were often firm. Their appearance was a brown or yellow color. When put to chemical tests, the stones transformed into powders when exposed to acids, and when they are examined by a chemist with the tool called the blowpipe, their color changes to black when exposed to fire and a distinct smell emerged from the first. They often made sounds as they broke apart in the fires. Marcet included detailed cross sections of several stones from the five major categories in the plates section of the book (Figs. 6.6, 6.7, and 6.8).

The stone categorized as the phosphate of lime was also defined as other chemists by another name [44]: "bone earth calculus." Phosphate of lime stones did not liquefy easily, and chemists had to add strong acids to break them down. When exposed to a chemical blowpipe, they turned colors, often black or white. Phosphate of lime stones could encase other stones as sheets. Made up of different types of salts and often forming dense, glittering minerals were classified as ammonia-magnesia phosphate. These stones produced a pungent ammonia smell when chemically analyzed, and the stones could be reduced into two distinct chemical divisions of phosphate, as well as magnesium. Other stones could grow into each other, and Marcet called these stones "fused." When fused stokes were put to chemical analysis, these stones left a powdery residue. These stones could be found in the bladder. Marcet had collected accounts where a soldier had been wounded by a gunshot to the bladder in the battle of Waterloo, and when the bullet was surgically excised, it had a fusible casing like a urinary stone around it.

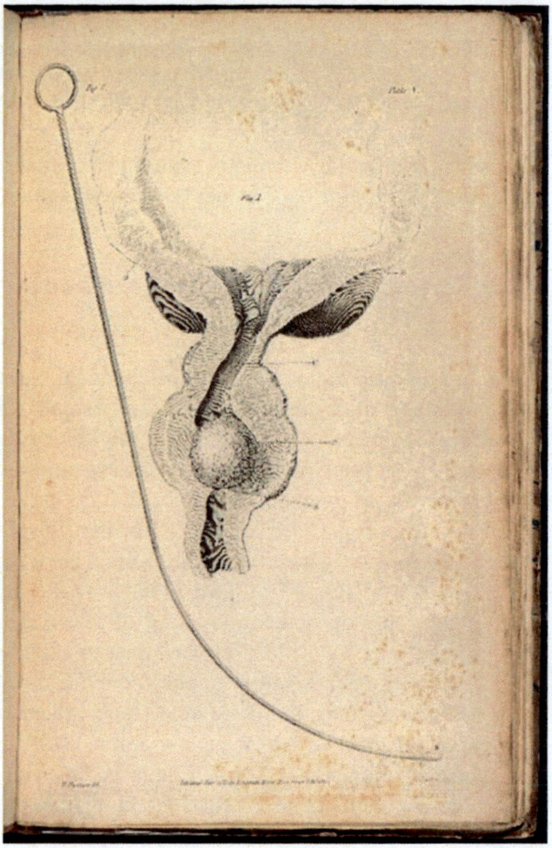

Fig. 6.6 Stone in urethra with file for grinding it (Science History Institute)

Another name for the oxalate of lime stone was a mulberry stone, as it resembled the fruit. When put to chemical analysis, especially heat, it could be reduced by the chemist to make "quicklime." It often had a dark color because it was often encased in the patient's blood, and the stones were often small in size, but Marcet needed more samples to test to get more information. Marcet then discussed the cystic oxyd calculi. It had a yellow color and was often somewhat see through when chemically analyzed and produced pungent smells when subjected to heat. It often burned like coal. Lots of acids destroyed this stone very easily. Much of the information included about this stone Marcet received from his surgical contributors.

Stones were often made up of many layers, and lots of stones had additional coverings of other stones on their surfaces. Marcet hoped that physicians and chemists could analyze stones to determine their base characteristics. He also examined previously underrepresented stones in the chemical literature, like prostate stones. Chemical analysis could reveal if stones actually originated in the

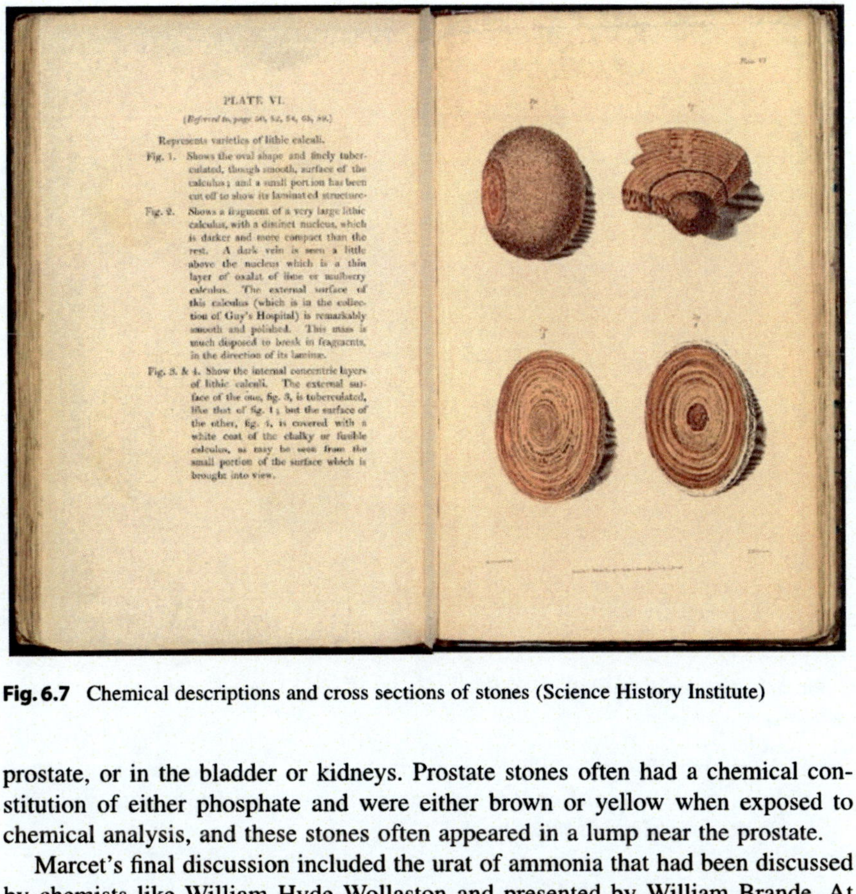

Fig. 6.7 Chemical descriptions and cross sections of stones (Science History Institute)

prostate, or in the bladder or kidneys. Prostate stones often had a chemical constitution of either phosphate and were either brown or yellow when exposed to chemical analysis, and these stones often appeared in a lump near the prostate.

Marcet's final discussion included the urat of ammonia that had been discussed by chemists like William Hyde Wollaston and presented by William Brande. At the Royal College of Surgeons, Marcet was an active member and had access to both surgical colleagues and the collection of urinary stones in the College's assortment of stones. Marcet mentioned remarkable stones in the collection, such as one from a gigantic snake, whom he classified as a [44] "Boa constrictor." He mentioned examining human stones and the stones found in organs in the collection, such as set of lungs. Many of the collections of human organs came from African-Europeans [45, 46].

Marcet wanted to reinstate that anyone could perform chemical analysis on urinary stones. Any doctor, chemist, or surgeon could contribute to the classification of stones [47], "It is not with pretension of offering anything new or important, to professed experimental chemists that I have introduced these details; but merely to enable those who may feel inclined to avail themselves to these hints, to select and procure, at the smallest apparatus necessary for carrying on experiment of this kind; and to obtain with great ease useful practical knowledge, upon a subject

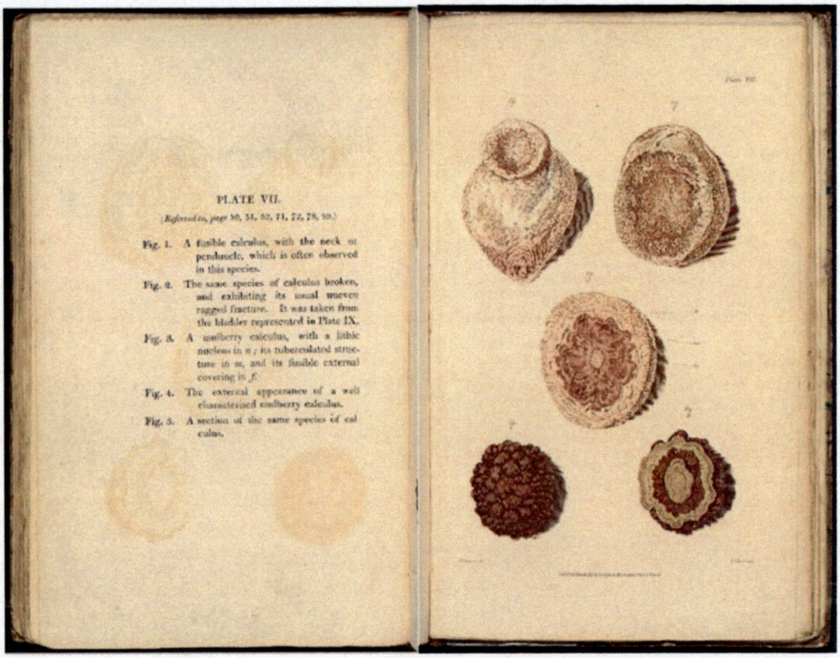

Fig. 6.8 Additional chemical descriptions of stones (Science History Institute)

which commonly supposed to present greatest difficulties and to require considerable chemical information." Anyone could learn chemistry of urinary stones by simply reading Marcet's short book.

6.6 Applying Chemistry to Treating Urinary Stones

Marcet wanted to discuss the potential for chemistry actually treating urinary stones, but he acknowledges that chemistry, no matter how effective, could never bring urinary stones to a complete end. Nonetheless, he includes the section "…The Chemical and Physiological Principles to be Attended to in the Treatment of Calculous Disorders." The best scenario for chemistry to the medical needs of patients suffering from urinary stones was that [48]:

> the only benefit which we may with any confidence expect from medicine in this disease, either to prevent the increase of calculi already formed, or what is still more important, to guard the constitution of those who are subject to the disorder against the prevalence of the particular diathesis from which it arises.

Surgery was still the best intervention for patients at the time. But he was still optimistic about chemistry being useful in either classifications or making the lithotomy less invasive to the patient [49]:

> At all events, since in attempting to remove calculi, we have to contend against unorganized bodies, which, though contained in living parts, do not obey the laws of the living principles, it may be fairly concluded, that, unless surgical aid be resorted to, it is in a great measure from principles that our views of treatment must be derived.

Chemical analysis of the urine was something that excited Marcet because of its potential to identifying stones and indicating the health of the patient.

Marcet needed to define what healthy urine was chemically, or do what the social scientists Canguilhem described as what was healthy and what was disease ("The Normal and the Pathological") [50]. Canguilhem, the teacher of Michael Foucault, argued that historical actors had to fully define both healthy states and diseased states [50]:

> The impetus behind every ontological theory of disease undoubtedly derives from therapeutic need. When we see in every sick man someone whose being has been augmented or diminished, we are somewhat reassured, for what a man has lot can be restored to him, and what has entered him can also leave. We can hope to conquer disease even if it is the result of a spell, or magic, or possession; we have only to remember that disease happens to man in order not to lose all hope...Disease enters and leaves man as through a door.

Consequently, Marcet needed to demarcate the healthy urine from the diseased urine so that he could do something for his patients and utilize all the chemical research he had performed. He defined urine that was normal or healthy as having an acidic quality to it, and he ran studies on normal urine. After having the sample remain on the bench for a day or two, it would produce chemicals like ammonia that was found then to have the ability to counteract acids which resulted in the production of salts. Marcet could also manipulate the urine sample to produce stone acid (lithic acid) with the addition of another agents, like vinegars or additional acids. He could also produce ammonia-magnesia-phosphate and also threw in limewater, which produced the building blocks of urinary stones that he discussed in the five principle categories. Marcet was interested to find that even health urine could be manipulated, presumably by the body's internal fluids, to produce stones, summarizing that [51], "whenever the lithic secretion predominates, the alkalies are the appropriate remedies; and the acids, particularly the muriatic, are the agents to be restored to, when the calcareous or magnesia salts prevail in the deposit."

Throughout his *Essay* Marcet continued to express excitement about the knowledge and potential end to suffering that chemistry could provide to urinary stone suffers, [52], "There is abundant evidence to prove that we are able in many instances to produce an effect sufficient to check the prevailing diathesis, and even sometimes to bring on a calculous deposit depending upon an opposite state of the system; a change which I have myself repeatedly witnessed." One day, Marcet likely felt that chemists could make changes in the body's fluids and keep acids from producing stones with the right chemical medicines for patients. Though he still did not think that acids could be taken into the body without danger, he did think some type of chemical remedies would be created that were safe enough for patients to consume. He built on the optimism because of the successful history of soda water at easing the complaints of patients suffering from the stone. Well-known chemists and physicians like Joseph Priestly and Dr. Benjamin Franklin promoted the use of soda water for the treatment of stones [53]. But Marcet did caution physicians and chemists of getting regular people to use too many chemical drugs, such as the popular use of magnesia patent medicines which seemed to further irritate the body. Most of the drugs that Marcet promoted were opioid based to treat patient pain or acids that seemed to open up blocked urinary passages [54]. He continued to investigate chemicals that could treat urinary stones throughout his career, but left the ending of his book as a hope to spur on additional research by chemists, surgeons, and physicians.

6.7 A Surgeon and Medico-Chemist Work Together

Marcet's collaborater, Sir Astley Cooper (Fig. 6.9), was originally from Norfolk. Cooper, born in 1768, had a bad reputation in his childhood as being undisciplined and harassing other students. However, he was shaken out of his adolescent interests and transformed by two major events in his life. The first event involved a surgeon effectively and efficiently treating a family member who had become injured. The second major event was his supreme curiosity in lithotomies after watching a surgeon successful treats a patient. Cooper then devoted his life to mastering surgery [55–58].[4] He started his training with the John Hunter, the famous anatomist, as well as William Cline. During his surgical training, Cooper became very ill and suffered from a fever during the year of 1787, and he traveled back to Edenborough to heal, where he continued to pursue anatomic education and chemical education. He made a full recovery in 1788 and returned back to his former teachers John Hunter and William Cline and eventually came a full member of then named Company of Surgeons, and their name would change in 1800 to the more professional sounding College of Surgeons [59].

While simultaneously practicing surgery, Cooper also taught anatomy as well. In 1791, he got married to Anne Cock, and as testament to his focus on surgery, Cooper spoke at anatomical lessons his wedding night. Cooper enjoyed traveling and was witness to many historical events, such as the French Revolution when he traveled to France in 1792. His life was not without tragedy, as his daughter Anna Marie, died in 1794, but he and his wife filled their house with two children through adoption and then added his nephew to his own through adoption as well. His nephew was named Astley in his honor. He had a steady position at Saint

[4] Cooper also has an excellent entry in the *Oxford Dictionary of National Biography*.

Fig. 6.9 Sir Astley Cooper (Wellcome Collection)

Thomas's Hospital teaching anatomy and performing surgeries. He met his friend Alexander Marcet at Guy's Hospital around 1800. Cooper and Marcet would be important to the start of the Medical and Chirurgical Society that was established in 1805.

Cooper's fame and position continued to increase with publications. Cooper's lectures on surgery and anatomy had become so popular that his lectures were published in their tenth edition by 1839. His work also appeared in mainstream medical periodicals, like the *Lancet*. By 1827, tragedy struck Cooper's life once again, and he had to retire from the practice of surgery for a period of morning, as his first wife died. In 1828, he had recovered enough to not only return to the practice of surgery but he remarried, and Catherine Jones became his wife. Cooper remained active in surgery, anatomic, and medical societies in London. Cooper died in 1841. As a teacher of anatomy, Cooper wished for his body to be dissected and the results of his autopsy published.[5]

Not only were Marcet and Cooper personal friends but they were colleagues that engaged with each other's publications, resulting in each of them referring to

[5] I have researched for the autopsy results and have so far not been able to procure them.

each other's works in the citations and references. These colleagues talked some interesting medical problems that found their way into the *Essay*. Cooper would often send Marcet stones that he removed surgically that he could not identify, and Marcet as a adapt chemist, would research and analyze them. The problem of the unknown stone would continue into the collections of the Royal College of Surgeons, which is discussed later in the book. Cooper did send descriptions with the stones, and Marcet included a chapter about two such mystery stones that did not fit into the five primary taxonomies of stones. Cooper's description of the first mystery stone was that of [60], "...a spherical calculus of the size of a large pea, to which were annexed the words [from Cooper], 'Is it cystic or uric?'" The color of the stone was that of yellow or brown, and the physical description of the stone was solid to the touch. It was also described by Marcet as "fibrous," which means it had a lot of fibers in its appearance. So Marcet called it fibrous as its own category. The stone also had a bad smell and did not act chemically like any of the calculi that Marcet had analyzed with his chemical tools.

Cooper also discusses his work with Marcet in publications, such as discussing the contributions of Marcet in his surgical notes. Marcet appears a character in many of his write ups of surgical cases meant for peer surgeons and students of surgery. And when Cooper discusses case descriptions of patients suffering from urinary stones, he explains how Marcet helped him, often with his chemical expertise [61], "A person came to consult me from the country with two openings, one above and one below the last rib, through which three calculi had been discharged. Dr. Marcet analyzed these, and found them to be composed of the ammoniac-magnesium phosphate." Cooper also echoes many of the trepidations Marcet described about the limits of chemistry [62]:

> With respect to the medical treatment of calculi, I do not believe in the power of chemistry to dissolve a stone in the bladder, if it acquires any considerable magnitude. The medicines, given for this purpose, become so much changed in their passage through the circulating and secreting system, that their chemical influence is in a great measure destroyed. They may alter the surface of a stone, so as to render it soft and less irritating; but they do not prevent a calculous secretion.

The two men shared the idea that chemistry had the biggest potential as a palliative and analytical tool for the provider.

Cooper also had chemical expertise of his own but knew that his expertise, like other providers had limits. He used a case summary where providers limited themselves to only chemical medicines and the providers did not try any other interventions [62], "I had a patient in Guy's Hospital with a stone in his bladder, in whom various experiments were tried to dissolve the stone by chemical menstrua. A catheter was introduced into the bladder, and through it injections were thrown; thus an opportunity was given for a direct application of the menstruum to the stone. After a lapse of time, I said to this man, 'Well, have my medical friends dissolved the stone?' His answer was, 'No, Sir, and I have given up all the injections except opium, from which I received considerable relief.' The patient died in the Hospital, and, on examination after death, a stone was found in his bladder." Cooper mentioned that chemical medicines had a palliative effect and recommended using soda or magnesium-based medicines and then recommended using acidic medicines to prevent the patient from suffering from further cases of stones. He had experiences where acidic medications prevent certain stones from returning [63], "the uric acid and oxalate of lime calculi return less frequently than the triple phosphate, which are very often reproduced."

Cooper would also mention Marcet in lectures, especially when discussing urinary stones. Cooper highlighted the contributions that Marcet brought to the surgical case in respect to chemistry. These cases were meant for peers and students to read, and perhaps, he was modeling the positive relationship chemistry could have with surgery. Cooper highlighted the contributions of Marcet in the pre-operation period of the case [64], "Admiral Douglas was the subject of stone; I sounded him, and in the evening of that day a portion of the stone was discharged by the urethra, and I sent it to my friend Dr. Marcet for analysis, who found it to be oxalate of lime; I therefore gave him acids, but he was not relieved by their use; he then took subcarbonate of soda ℥ss four times in the day, in some water." Though Cooper thought that chemical analysis was important to surgical cases, he did not envision chemistry replacing surgery, and he recounted the difficulties of chemists in actually dissolving stones in the body [64],

> Some months afterwards I was requested to meet Dr. Reynolds and Sir E. Home in consultation upon the case of the Duke of Portland; and when I entered the room, Sir Everard said, "Cooper how did you dissolve the stone in Admiral Douglas?" to which I answered, "I never dissolved a stone in my life."—"But" said Sir Everard, "he expressed himself well from some medicine you ordered him." I called in consequence on the Admiral at his hotel; when he said, "You saw me in dreadful agony, unable to cross a room; but since I have taken the soda, I went from Yarmouth, in Norfolk, to Portsmouth, by land, and bore the journey well; and I could now go down a country dance." Yet the stone still existed in his bladder; but the soda had lessened its sensibility, so as to enable him to bear the complained without much suffering, and only a little inconvenience from the stone, which still occasionally stopped the flow of urine.

Many of the surgeons mentioned in this case study also participated in the Medico-Chirurgical Society, which featured discussions of the chemistry of urinary stones, and they also likely knew Marcet. It is no surprise that they echoed many of the skepticisms and cautions Marcet discussed in his *Essay*.

6.8 Humoral Pathology in the Surgical Mind of Astley Cooper

Irritations were also in the forefront of surgical minds like Cooper. Irritations affected the chemistry of the body. Some surgeons thought too much bodily irritation in operations, like lithotomies, caused patients to die [65]. Many of the surgical cases that Cooper discussed in his *Lectures* discuss humoral pathological ideas beyond irritability and also include inflammation, blood hemorrhage, and diseases of the organs that are affected by fluid problems [65]. Irritation could affect the urinary organs, tissues around the body, kidneys, and areas around the bladder. Irritability that originated from nervousness could cause the whole body to become overwhelmed and kill the patient, such as cases of children who had weaker constitutions than adults. Surgeons only tool in treating this type of irritability was opioid based or other drugs that calmed the system or medicines that removed the irritation from the system, such as pukes. He also called for the continuation of very humoral-based tools like blisters and leeches in order to remove irritation from the body. Cooper also called for bleeding and baths to relieve irritation in the system and restore harmony in the fluids of the body [66].

One situation that stressed Cooper was patients bleeding out too much. These bleeding conditions "repeatedly destroy life," but it was poorly understood and often not attributed to the death of a patient [66]. Cooper disagreed with doctors like Benjamin Rush, who advocated intensive bleedings, and Cooper thought that it should be done more judiciously [65], "…should not quit his patient until the bleeding caused by the operation has ceased: the patient should not be put to bed whilst any hemorrhage continues; and when in bed he should be very lightly covered for some time."

Constitutional issues also contributed to patient suffering and death. Coper speculated that when the patient's scrotum turns gangrene, it was because of either a weakness in their constitution or blamed it on some type of lifestyle issue. Urine itself led to irritation of the scrotum, and if the urine was unhealthy, it caused continual issues for the patient by not only causing scrotum problems but problems in the tissues of the urinary organs. The patient could also face intensive irritations from surgeries that caused organ failure, such as in the liver, heart, lungs, or quicken the pulse in the body that led to death. He described irritation in the context of organ failure causing the [67], "…morbid state of the liver; dyspnea from some chronic affection of the lungs; palpitation of the heart; irregular or intermitting pulse; which tend to destroy the powers of restoration." The liver was an important organ for humoral pathologists, as they believed it played a major role in keeping the body alive. Historically, ancient physicians and early practitioners of humoral medicine saw the liver as the place where the body's vital energy was circulated through the body [68].

Constitutionally, Cooper saw women as a different type of patient and perhaps was trying to explain the different outcomes in regards for the lithotomy that Marcet had discovered. Cooper shared the same ideas about women that Marcet discovered, that in the future they might not need lithotomies, and Cooper singled out discussions of women in his *Lectures* as special patients. He historicized his ideas about women and lithotomies [69], "Lithotomy is much less frequently required in the female than in the male, probably on account of the meatus readily permitting the escape of materials which would have become the nuclei of stones in the male, be they portions of gravel, of blood, inspissated mucus, or extraneous bodies." But Cooper believed that because of constitutional differences, women experienced pain from lithotomies, perhaps more than men, because of issues in their lifestyle and therefore were, "subjects of lithotomy from perverse and unnatural propensities." He noted that some women chose to put rocks in their urethras, presumably as a symptom of mental illness to intentionally invoke the need for a lithotomy [70]. Cooper discussed female patients suffering from mental illness trying to get lithotomies on purpose [71], "I have known women introduce extraneous substances into the vagina, to invite the operation for the stone." But he offered up no examples nor case histories that seemed to fit his speculation.

Cooper continued to echo Marcet and other medico-chemists cautions about imbibing acids as therapies [72], "it might be thought that solvents could with advantage be injected, but the patients cannot bear them, and will not submit to their use, as they irritate excessively." The patient's constitutional makeup was also important, as he believed that urinary problems could cause intense constitutional problems for women. Cooper cautioned that women could suffer from hysteria if there was too much urine retained in their bodies [73]. Stones could also cause problems in the uterus and further cause sympathetic diseases, like hysteria, or blocked fluids could reap havoc in a woman's body.

Stones could also appear in other areas of the body, such as in the cheeks where saliva was produced [74]. A stone could appear where saliva was produced and block the duct, and he reported that patients found these stones distinctively excruciating. Cooper reported the work of another Medico-Chirurgical Society member, surgeon Henry Cline, who suffered from salivary stones. Cline experienced a salivary stone while living with Cooper, and he reported how Cline suffered [74],

> …[H]e used frequently to say, 'I have a spasm in my mylohyoideus muscle,' and it was usually at the time of eating that he made this observation: at length he said, 'I have discovered the cause of the uneaseiness [sic?] and spasm under by tongue, it arises from a stone in the submaxillary duct,' which he desired me to feel, and which I removed from him….

Different airs could provide relief from salivary stones, such as living near the beach. Cooper also advised salt water treatments, but surgery was often the most effective form of relief, as Cooper reported that these stones could grow as massive as a common almond, but these large stones were removed through surgery. The surgery was especially painful because it often involved the surgeons placing books into the duct of the patient, located at their cheek, and other apprentice surgeons pushed on the duct to force the stone out. When that did not work, the stone was expurgated by the lancet and destroyed below the patient's tongue, exposing access to additional blocked ducts. Stones were also often removed with the large lithotomy forceps (Figs. 6.8 and 6.10).

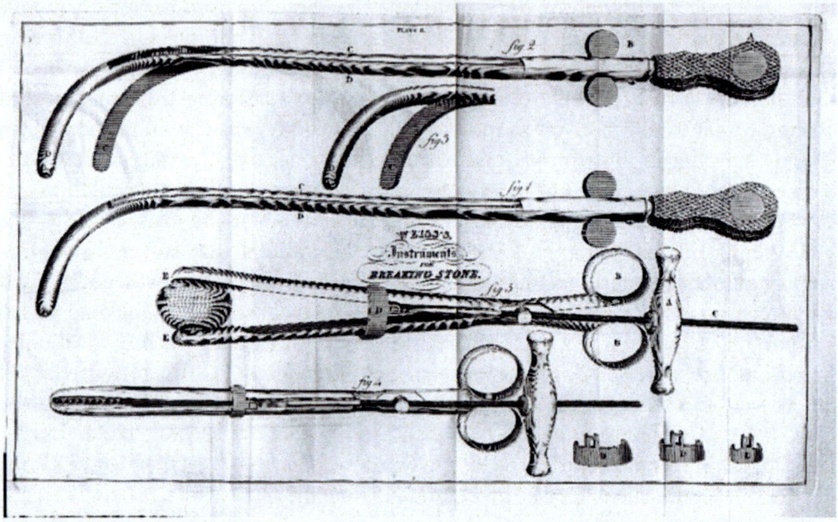

Fig. 6.10 Surgical tools for breaking stones, from John Weiss, *An account of inventions and improvements in surgical instruments made by John Weiss, 62, Strand…* (1831), Reynolds-Finley Historical Library, the University of Alabama at Birmingham, with permission

6.9 Boundary Objects and Partners in Friendship, Medicine, and Surgery

Urinary stones brought friends like Alexander Marcet and Astley Cooper together because of their interests in understand diseases related to the fluids of the body, but also their interests in improving surgery with better chemical knowledge. The boundary object that brought not only these two people together but brought the fields of chemical medicine and surgery together were the stones of the body. Alexander Marcet wanted to improve chemical knowledge in order to better classify urinary stones and expand chemical testing of stones. He needed access to collections of urinary stones and other bodily stones, from surgeons who removed them like Astley Cooper. Cooper wanted to improve surgical practice and his own knowledge about chemistry, as he was an intellectually curious surgeon. The two men worked together, and the results were communicated in Marcet's *Essay* in terms of chemical classifications of stones, statistics about lithotomies, and better knowledge about the numbers of patients that suffered from stones. Cooper benefited from reading and learning from Marcet's chemical work, as he was able to communicate the latest chemical ideas to surgical students through his own publications. Their collaborations did not receive much disciplinary objections and likely reflected the unhindered intellectual questions in the health sciences in during the nineteenth century [75, 76]. Cooper and Marcet were also able to communicate cautions about the limits of chemical knowledge and remedies in their respective publications and audiences. Their collaborations, partnerships, and friends were mirrored in intellectual societies in the nineteenth century that were also interested in questions related to improving knowledge about bodily stones and the fluids of the body.

References

1. Leigh JG (2018) Alexander Marcet, chemist, physician and geologist, a neglected figure in British Science from 1797 to 1822. Bull Hist Chem 43:1–13
2. Coley NG (1968) Alexander Marcet (1770–1822), physician and animal chemist. Med Hist 12:394–402
3. Marcet A (1817) An essay on the chemical history and medical treatment of calculus disorders. Longman, Hurst, Rees, Orme, and Brown, London
4. National Library of Medicine (2022) Greek Medicine: https://www.nlm.nih.gov/hmd/greek/greek_oath.html
5. McPherson L (2015) The history of the Hippocratic oath. Northeastern Nursing Blog: https://absn.northeastern.edu/blog/the-history-of-the-hippocratic-oath/
6. Indla V, Radhika MS (2019) Hippocratic oath: losing relevance in today's world? Indian J Pysch 61(Supp 4):S773–S775
7. Ellis H (2020) The story of 'cutting for the stone'—lithotomy for bladder calculus. J Perioper Pract 30:183–185
8. Riches E (1968) The history of lithotomy and lithotrity. Ann R Coll Surg Engl 43:185–199. Citation found in Bibliography of the history of Medicine 1964–1969 by U.S. Department of Health, Education, and Welfare
9. Ellis H (1979) A history of bladder stone. J R Soc Med 72:248–251
10. Carpue JC (1818) A history of the high operation for the stone. Longman, Hurst, Rees, Orme, and Brown, Paternoster Row, London
11. Star SL, Griesemer JR (1989) Institutional ecology, 'translations' and boundary objects: amateurs and professionals in Berkeley's Museum of Vertebrate Zoology, 1907–39. Soc Stud Sci 19:388–389
12. Weisz G (2006) Divide and Conquer: a comparative history of medical specialization. Oxford University Press, New York
13. Bold AM (1965) Use and abuse of clinical chemistry in surgery. BMJ 2:1051–1052
14. Bynum W, Porter R (eds) (1985) William Hunter and the eighteenth-century medical world. Cambridge University Press, Cambridge
15. Bynum W, Porter R (eds) (1985) William Hunter and the eighteenth-century medical world. Cambridge University Press, Cambridge, p 3
16. Starr P (1982) The social transformation of American medicine. Basic Books, New York
17. Fothstein WG (1985) American physicians in the nineteenth century: from sects to science. The Johns Hopkins University Press, Baltimore
18. Wilson P (1995) Enlightenment science? Surgery and the royal society. In Porter R (ed) Medicine in the enlightenment. Rodopi, Atlanta, pp 360–386
19. Fanfan D, Ehrlich H, Elkuli A (2021) The future of the surgeon-scientist: a journey funneled through inspiration, roadblocks, and resilience. Ann Med Surg 62:65–67
20. Rosenfield L (2001) The chemical work of Alexander and James Marcet. Clin Chem 47:784–792
21. Marcet A (1823) The annual biography and obituary, for the year 1823. Longman, Hurst, Rees, Orme, and Brown, London
22. Whatmore R (2012) Against war and Empire: Geneva, Britain, and France in the eighteenth century. Yale University Press, New Haven

23. Palmer RR (1959) Age of the democratic revolution: a political history of Europe and America, 1760–1800, vol 1. Princeton University Press, Princeton
24. Dann O, Dinwiddy J (1988) Nationalism in the age of the French Revolution. The Hambledon Press, London
25. Rothenberg GE (1988) The origins, causes, and extension of the wars of the French Revolution and Napoleon. J Interdiscip Hist 18:771–793
26. (1823) Alexander Marcet. The annual biography and obituary, for the year 1823. Longman, Hurst, Rees, Orme, and Brown, London, p 291
27. Bahar S (2001) Jane Marcet and the limits to public science. BJHS 34:29–49
28. Polkkinghorn B (1982) Political economy disguised as fanciful fables. East Econ J 8:145–156
29. Thomas JM (1991) Michael Faraday and the Royal Institution: the genius of man and place. Taylor & Francis, New York
30. (1823) Alexander Marcet. The annual biography and obituary, for the year 1823. Longman, Hurst, Rees, Orme, and Brown, London, p 296
31. Hunting P (2002) The history of the Royal Society of Medicine. The Royal Society of Medicine, London
32. Marcet A (1817) An essay on the chemical history and medical treatment of calculus disorders. Longman, Hurst, Rees, Orme, and Brown, London, pp viii–x
33. Marcet A (1817) An essay on the chemical history and medical treatment of calculus disorders. Longman, Hurst, Rees, Orme, and Brown, London, pp xi–xiii
34. Marcet A (1817) An essay on the chemical history and medical treatment of calculus disorders. Longman, Hurst, Rees, Orme, and Brown, London, p xii
35. Marcet A (1817) An essay on the chemical history and medical treatment of calculus disorders. Longman, Hurst, Rees, Orme, and Brown, London, p ix
36. Marcet A (1817) An essay on the chemical history and medical treatment of calculus disorders. Longman, Hurst, Rees, Orme, and Brown, London, p 1
37. Marcet A (1817) An essay on the chemical history and medical treatment of calculus disorders. Longman, Hurst, Rees, Orme, and Brown, London, p 2
38. Marcet A (1817) An essay on the chemical history and medical treatment of calculus disorders. Longman, Hurst, Rees, Orme, and Brown, London, p 7
39. Marcet A (1817) An essay on the chemical history and medical treatment of calculus disorders. Longman, Hurst, Rees, Orme, and Brown, London, p 8
40. Driggers EA (2021) "The voice of the stomach": the mind, hypochdriasis and theories of dyspepsia in the nineteenth century. Hist Pysc 32:58–99
41. Frangsmyr T, Heilbron JL, Rider RE (eds) (1990) The quantifying spirit in the eighteenth century. University of California Press, Berkeley
42. Marcet A (1817) An essay on the chemical history and medical treatment of calculus disorders. Longman, Hurst, Rees, Orme, and Brown, London, pp 25–27
43. Marcet A (1817) An essay on the chemical history and medical treatment of calculus disorders. Longman, Hurst, Rees, Orme, and Brown, London, p 26
44. Marcet A (1817) An essay on the chemical history and medical treatment of calculus disorders. Longman, Hurst, Rees, Orme, and Brown, London, p 63
45. Driggers EA (2019) The chemistry of blackness: Benjamin Rush, Thomas Jefferson, Everard Home, and the project of defining blackness through chemical explanations. Crit Philos Race 7:372–391
46. Otele O (2021) African Europeans: an untold history. Basic Books, New York
47. Marcet A (1817) An essay on the chemical history and medical treatment of calculus disorders. Longman, Hurst, Rees, Orme, and Brown, London, pp 122–123
48. Marcet A (1817) An essay on the chemical history and medical treatment of calculus disorders. Longman, Hurst, Rees, Orme, and Brown, London, p 143
49. Marcet A (1817) An essay on the chemical history and medical treatment of calculus disorders. Longman, Hurst, Rees, Orme, and Brown, London, p 144
50. Canguilhem C (1989) The normal and the pathological. Zone Books, New York, p 39

51. Marcet A (1817) An essay on the chemical history and medical treatment of calculus disorders. Longman, Hurst, Rees, Orme, and Brown, London, p 148
52. Marcet A (1817) An essay on the chemical history and medical treatment of calculus disorders. Longman, Hurst, Rees, Orme, and Brown, London, p 150
53. Finger S (2006) Doctor Franklin's Medicine. University of Pennsylvania Press, Philadelphia
54. Lomax E (1973) The uses and abuses of opiates in nineteenth-century England. Bull Hist Med 47:167–176
55. Cooper BB (1843) The Life of Sir Astley Cooper, Bart: interspersed with sketches from his note-books of distinguished contemporary characters, vols 1–2. John W. Parker, London
56. Cooper A (1824–1827) The lectures of Sir Astley Cooper on the principles and practices of surgery. Thomas and George Underwood, London
57. Moore W (2006) The Knife Man: blood, body snatching, and the birth of modern surgery. Broadway Books, New York
58. Lawrence SC (1996) Charitable knowledge: hospital pupils and practitioners in eighteenth-century London. Cambridge University Press, Cambridge
59. Hewlett RT (1906) The Royal Medical and Chirurgical Society of London Centenary 1805–1905. Nature 73:388
60. Marcet A (1817) An essay on the chemical history and medical treatment of calculus disorders. Longman, Hurst, Rees, Orme, and Brown, London, p 110
61. Cooper A (1824–1827) The lectures of Sir Astley Cooper on the principles and practice of surgery, vol II. Thomas and George Underwood, London, p 168
62. Cooper A (1824–1827) The lectures of Sir Astley Cooper on the principles and practice of surgery, vol II. Thomas and George Underwood, London, p 181
63. Cooper A (1824–1827) The lectures of Sir Astley Cooper on the principles and practice of surgery, vol II. Thomas and George Underwood, London, p 183
64. Cooper A (1824–1827) The Lectures of Sir Astley Cooper on the principles and practice of surgery, vol II. Thomas and George Underwood, London, p 182
65. Cooper A (1824–1827) The lectures of Sir Astley Cooper on the principles and practice of surgery, vol II. Thomas and George Underwood, London, p 200
66. Cooper A (1824–1827) The lectures of Sir Astley Cooper on the principles and practice of surgery, vol II. Thomas and George Underwood, London, p 201
67. Cooper A (1824–1827) The lectures of Sir Astley Cooper on the principles and practice of surgery, vol II. Thomas and George Underwood, London, p 202
68. Arikha N (2007) Passions and Tempers: a history of the humours. Ecco, New York
69. Cooper A (1824–1827) The lectures of Sir Astley Cooper on the principles and practice of surgery, vol II. Thomas and George Underwood, London, p 224
70. Ross S (2018) Pathology and performance: the female body in the romantic era. Ph.D. dissertation, University of Arizona, Tucson, p 19
71. Cooper A (1824–1827) The lectures of Sir Astley Cooper on the Principles and Practice of Surgery, vol II. Thomas and George Underwood, London, p 225
72. Cooper A (1824–1827) The lectures of Sir Astley Cooper on the principles and practice of surgery, vol II. Thomas and George Underwood, London, p 226
73. Cooper A (1824–1827) The lectures of Sir Astley Cooper on the principles and practice of surgery, vol II. Thomas and George Underwood, London, p 230
74. Cooper A (1824–1827) The lectures of Sir Astley Cooper on the principles and practice of surgery, vol II. Thomas and George Underwood, London, p 228
75. Makari G (2009) On the shifting boundaries of medicine. Lancet 373:P206-209
76. Estes JW (1988) The pharmacology of nineteenth-century patent medicines. Pharm Hist 30:3–18

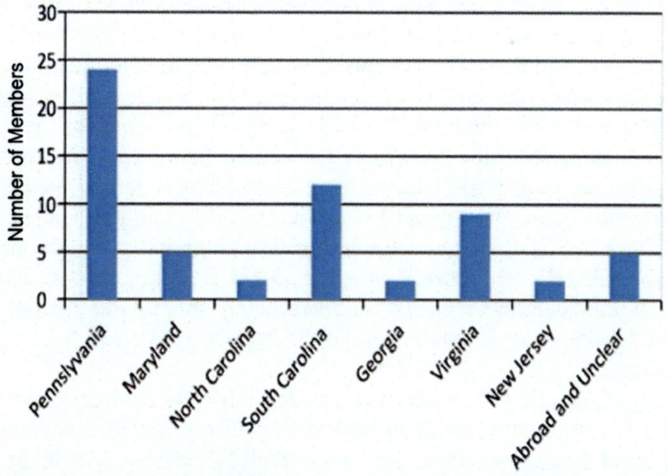

Fig. 7.1 States where members originated

graph given in Fig. 7.1 represents membership by states and offers a comparison to membership outside of the United States.

The graph includes members, as well as those members who were made in honorarium by the society. The graph was compiled by examining the named home area on medical dissertations completed at the University of Pennsylvania. Dissertations of students often included recognition of membership in medical societies, as well as scientific societies like the Chemical Society of Philadelphia. With digitization efforts performed recently at the University of Pennsylvania, these numbers could increase, as many of the dissertations examined came from the digital version of the *Early American Imprints Series I-II*.[2] The nature of student participation in the Chemical Society of Philadelphia is also unclear. Further complicating the narrative of the Chemical Society of Philadelphia is its short duration, as there are few formal records left about the society.

From the data that was able to be collected, some trends in membership can be noted. First, in analyzing membership by American state, Pennsylvania had the highest, as a lot of Pennsylvania students attended the University of Pennsylvania because of its locality [14]. It is also interesting to note that the second most popular state for membership in the Chemical Society of Philadelphia is a southern state, South Carolina. In comparison, it would take around 720 miles to travel

[2] Inaugural dissertations available on *Early American Imprints: Series I-II* constituted the estimation of membership. Most students mentioned their affiliation on the title page of the dissertation. However, some dissertations lacked information about the student's home state, or they omitted it. I have a list of members identified through dissertations that I would be more than happy to share.

from Charleston to Philadelphia on horse or carriage, which I estimated in driving miles. It seems that Northern and Southern medical students could travel with somewhat ease, and Southern students often left the South to attend institutions of higher education in the nineteenth century [15]. In comparing the regions of the United States, it is important to try and generally define them for the international reader, though establishing regions in the United States can be a historiographically perilous activity [16]. The Northern United States can be generally defined with states like Pennsylvania and Massachusetts. The Southern United States consists of states like Maryland and South Carolina. If these two regions are compared by membership, slightly more Southern students were members of the Chemical Society of Philadelphia, with the numbers being 48% of the membership coming from Northern states, while 52% of the membership came from the Southern United States.[3]

Writing the history of an organization that left little records is a difficult task, and there are some disagreements with the small number of historians who have written about the society. Generally, a medical doctor served as its first and only president. James Woodhouse was a famous member of the early Philadelphia medical community [17]. Woodhouse lived from 1770–1809, and it was early in his career when he served as president. The participation of students in the society's activities or organization is unclear. Woodhouse's career is a model for interdisciplinary training, as he was both a surgeon and a chemist. He started his career on the Western frontiers of America, serving as a field surgeon to American forces fighting American Indians [18, 19]. After shifting away from surgery to chemistry, he later taught chemistry at the University of Pennsylvania. During his time educating students in chemistry from 1795–1809, Woodhouse encouraged passionate discussions of chemical ideas, and he engaged in debates with prominent chemists, like Joseph Priestly, over the nature of phlogiston. Phlogiston was the substance that was thought to emerge from burning substances prior to the theory of oxygen combustion associated with Lavoisier, but some historians would argue that should be associated with Priestley, or even Carl Wilhelm Shceele [20–23].[4] During Woodhouse's time as president, he promoted the importance of chemistry in medication education, much like the work of Thomas Cooper in a previous chapter in this book. Woodhouse gave a speech promoting the importance of chemistry to medicine.

The Philadelphia Gazette published a text of the speech that Woodhouse gave in front of society membership on June 21, 1797 [8, 17, 24]. Woodhouse added swipes at the religious establishment as well, as he felt that the repression of the Church was hurting scientific production, chemists like Joseph Priestley, and

---

[3] At some future point I plan on writing up a follow up article about what happens to the members of the society in regards to their professional trajectory. I have a full list of members and am continuing to add to it.

[4] The nature of the discovery of oxygen is a major point of debate in the history of chemistry. Please see the citations referred to in the text. A really good place to the start is Kuhn's discussion in *The Structure of Scientific Revolutions*.

another chemist, Samuel Coates, to be stimmed by clergymen, or "old men" who did not appreciate the new and exciting ideas from chemistry and of the youth who were open to new scientific concepts. Rebellion against conservative institutions and religion were some of the "Secrets" that Woodhouse claimed he was imparting to his audience.

Another deterrent to students pursuing medicine and chemistry in Philadelphia were physicians who did not agree that chemistry was useful in the practice of medicine. Skeptical physicians of Philadelphia cautioned students from pursuing chemistry [24], that "some of the Physicians of thi[s?] [sic] city, who are persuading you from studying chemistry, have the appetite without the wing, others have neither the wing nor the appetite. You, gentlemen, have wings, and the old men would cut theirs off, if they could." He romanticized the notion of having a passion for chemistry, analogizing it to dating a woman. Woodhouse further feminized chemistry [24], "I have found out a sweetheart for you all. Court miss Chemistry; she is a young lady, professed of an amiable disposition. Rival each other in attention to her, and she will crown you with laurels. Marry her as soon as possible, she will not quarrel with you a few months after entering to the conjugal state. You will never divorce her, and the honeymoon will last for life. I have set you the example, for I love this lady to distraction. Let every individual of our society then, despite the contemptible passion of love, and cry out, Hail! Ambition, thou art my god."

He equally chastised physicians who were not interested in his chemical program. He thought that the reputation of the United States resisted on the success of chemical education and found little logical reasons for students not pursuing chemical education [24]. He also compared pharmacy to pursuing a romantic relationship with a woman [24], "The member of the college has told us, that his brethren have conceived a design of publishing a Pharmacopoeia. I can inform him, that the brat has been in the womb ten years, and by this time has certainly perished."

Woodhouse saw a destiny for the Chemical Society of Philadelphia, and it was destined to provide greatness for the newly established United States. He saw the importance of his chemical society in the context of American nationalism [24], "if we do not succeed in becoming the fathers of chemistry in the United States of America, I will like Paracelsus, retire to a cave, weary of the vices of follies of mankind." The legacy of the society is demonstrated in the number of students who were interested in both medicine and chemistry in their education. Moreover, many of the members of the society continued to work in medicine and chemistry, and their published research reveals the interests of many of these members in the stones of the body and humoral pathology.

One member, Felix Pascalis, was a member from outside of the United States, originating from France [25]. Pascalis was originally from Provence and studied medicine in the Montpellier area of France. He later left France and traveled to modern day Haiti, which was called St. Domingo prior to the Haitian Revolution [26]. Pascalis likely traveled to the profitable sugar island to practice medicine for the planters and enslaved workers [27]. When the Revolution turned violent and prior to the French invasion of the island, Pascalis left, like many French aristocrats, to immigrate to Philadelphia, as Philadelphia became a refuge for immigrants not only from revolution in the Caribbean, but in the old world, as aristocrats left France when the Revolution became violent there as well [28]. Pascalis lived in Philadelphia for a time, and then moved to New York, where his life ended in 1833 [25]. During his life, Pascalis was active in creating medical knowledge and publishing medical periodicals, as he served as one of the editors of the journal *Medical Repository*. He also published books on fevers and yellow fever. Pascalis was interested in medico-chemistry and also published work on the chemistry of fevers. Other publications include writing about his own health, as he wrote about his own case of liver disease. The case histories he published also included his attending physicians that included Charles Caldwell, Benjamin Rush, and Phillip Syng Physic [29]. Pascalis was also interested in southern medical communities, and was a member by correspondence, with the South Carolina Medical Society [30, 31].

As a member of the Chemical Society of Philadelphia, Pascalis decided to give the yearly public speech for the society in 1802 [32]. After discussing the importance of chemistry to understanding the flora, fauna, and rocks of the earth, he turned to the promotion of chemistry in cultivating the power of medicine. Prior to the addition of chemistry to medicine, humoral pathology was in a lot of trouble with competing ideas of medicine [32]. It was unclear if this was a derogatory remark at the humoral medicine of the past, but he did promote the importance of understanding how the body's many parts functioned, and this reductionism resembled the improvements of chemistry. He believed, like many individuals in this book that chemistry was crucial to making medicine successful [33]:

> at last, in imitation of chemistry, the spirit of analysis has prevailed in all the branches of natural philosophy, and consequently the friends of the healing art, who wished independently and usefully to pursue the career of their labour, have renounced all logical systems, and composed their institutes of medicine of such facts of aphorisms of the ancient and modern, that experience had rendered inconvertible with all the results that physical laws and analytic animal chemistry could consistently offer to their medical investigations.

He wanted his audience to be cautious of theories that combined physiology and chemistry without a lot of careful thought, as these new ideas could be "...mistaken for physiological and Chemico-medical improvements" [33]. But understanding animal chemistry clearly improved medicine to Pascalis and he was interested in chemists who were researching the chemistry of life [34]:

> I do not pretend to controvert here the supposed discoveries of Girtanner, of Beddoes, Davis and others, who with so rapid strides through the scabrous path of science, have promised to themselves and the public, to open a new Æra, and to dispel from among making all the diseases they were necessarily subject to.

Pascalis continued to fan the optimism that chemistry was a panacea to healing. He promoted the fact that new and path breaking discoveries were made all the time in chemistry, and they would likely contribute to medicine [35]. The gases and fevers of life were also under the purview of chemical analysis [35], "animal acids chiefly, and other primarily combinations in the blood, in the bile, in the bones, in earthy concretions and others, do form, Gentlemen, the most precious collection of facts and observations, that ever medical science could be improved with, for the relief and cure of a great many diseases." Chemistry was also a tool that would legitimize pharmacy and chemistry itself was legitimated by its mass spread into colleges and universities around the United States [36]. The speech was an advertisement for scholars and teachers to continue to engage with chemistry and find continual usage for its methods.

As a researcher, Pascalis wanted to reduce diseases and bodily illness down to its chemical components. Pascalis was similar to Benjamin Rush as a chemist in that manner, especially when he engages with race as a medical category. Pascalis examined the question of how people's skin could suddenly change color in an 1818 article that appeared in *Medical Repository* [37, 38]. There had been other European authorities who had published cases where patients or individuals had suddenly changed their skin color, like Gottfried Van Swieten, and also noted that there had been other cases where white people had their skin color change temporarily. Pascalis' worldview saw skin color as changeable [39], "this fact seems to have researched an equal change to that of an opposite change from black to white in the human species." One example that Pascalis cited was that of Mary Gillard, a French woman who had changed from having white skin to black skin temporarily. She was around seventy years of age and had become impoverished and lived her life homeless, and had to beg for food. She spent her days caring for children of her daughter, while Gillard's daughter worked, but she had to struggle as her children became ill. Tensions arose between Gillard and her daughter as her daughter blamed their misfortunes on her, and also put their illnesses on Gillard as well. The daughter even claimed that Gillard gave her grandchildren syphilis and it was clear that Gillard was under a lot of stress. Pascalis summed up the situation [39], "These events led the mother to a deep grief. The daughter threw herself out of a window, with her two children in her arms, and died. Mary Gillard then turned black from her head to her feet."

Gillard was taken into the hospital, but doctors had no idea what were causing these changes in her body. Gillard had her face, hands, groin, and feet covered in dark or black skin, and she also developed black spots on her arms, but her legs did not turn black. The patient was in intense pain and fear [39]. Treating the irritation in Gillard's body, a blister was administered, which opened a huge wound in the body, which released fluids. Pascalis thought that the skin of Gillard reacted in the same way that a patient with dark skin would react to a blister [39], "...the cuticle, and the cellular tissue were all blackish, as it happens on the skin of the negroes." Pascalis noted that the fluids were now flowing freer and the body was in a better position of health because of the interventions, such as the use of the lancet and blister to pull bodily fluids outside of the skin. Unfortunately, the patient died and a postmortem was performed by Pascalis, where he found that there was a problem with Gillard's fluids that he described as having the character of [40], "...a considerable effusion of bilious and purulent matter."

After the case history of the patient, Pascalis continued to speculate about what happened internally in the patient's body to cause the skin changes. He speculated that it was foremost a change in the body's fluids. There was a fluid that was responsible for black skin, much like Rush's theories about the leprosy disease that caused black skin, it was a disease that was based in the fluids of the body. Pascalis speculated that fluids were the cause of Gillard's disease [40], "that the secretion of *black matter* exists in the white race, can be proved by numerous facts in the heathy as well as in the morbid condition of the same." Pascalis framed black skin color as pathological and diseased, while white was normal and healthy, much like Benjamin Rush.

Skin color, according to Pascalis's ideas, presented in the womb and every human was born with sometime of fluids that turned parts of the body black. White people had those fluids in their eyes and even in their hair. The black liquids could also be removed from the body through purgatives. Pascalis believed that this black fluid was the origin of sickness, which caused fevers and stomach ailments. He even thought that the black fluids of the body might be the reason for the appearance of yellow fever [41]:

> Medical practitioners have sometimes supposed that certain medicines, especially calomel could effect a change of colour in the alvine secretions, and of bile. But in the yellow fever this secretion becomes fatal, and is the most universal and dreaded symptom, for it is produced by it in such a quantity as to astonish, and to bespeak the dissolution of the whole blood.

Black fluids in the body might cause other diseases from consumption to breathing disorders, and generally cause irritation in the body such as [41], "...pulmonary disease or consumption, forming tubercles in the substances of the lungs, that are black and carbonaceous." Though Pascalis did not frame diseases in an explicitly humoral way, he thought that the black liquids in the body caused disease in the same way that yellow liquids also caused disease, much like black and yellow bile in ancient humoral theory. Pascalis, though he was using terms from chemistry, was using the same structure as humoral theory in framing disease in the body.

He further speculated how blood was important in the disease processes in the body, especially when there was a problem based in the black fluid in the body. Patients of African descent had an excess of black fluids in the body, as Pascalis thought that the air caused the build-up of black fluids. Pascalis thought that large amounts of carbon caused white skin, while African people had better blood systems that took in and accepted carbon more efficiently, and this carbon was removed through breathing the warm airs in Africa. Carbon was stored in the skin areas of African people, but any type of stress could cause that carbon to remove itself from the body. Pascalis saw diseases in the body as highly active and sensitive to irritations [42]:

> by violent causes, or a morbid alteration of organization, the black matter may disappear from the skin of an African, and from carbonic acid gas by respiration; or vice versa, the carbon of the blood in a white person may be wholly deposited on the skin, as in the case of the unfortunate *Mary Gillard*, and others of the same nature, whose lungs have been by some cause prevented from secreting the carbon of the venous blood.

The body's fluids were highly susceptible to changes in stress levels, the airs, or other external changes, and those internal irritations had external markers. The body's chemistry reflected the body's internal disease processes.

Members of the Chemical Society of Philadelphia like Pascalis were highly interested in the chemistry of life and how beneficial chemistry could be to the improvement of medicine. Societies like the Chemical Society of Philadelphia also researched bodily diseases and framed these diseases as caused by internal fluid obstructions. Chemical research into stones, diseases, and fluid blockages continued during the 1800s with the Columbian Chemical Society.

7.1.2 Columbian Chemical Society of Philadelphia

Physicians in Philadelphia founded another society focused on chemical research in 1811 [43–45]. Students of Benjamin Rush who had received extensive education founded the Columbian Chemical Society of Philadelphia to continue discussions and debates about chemistry. Thomas D. Mitchell, as well as George F. Lehman, established the Columbian Chemical Society, who had both studied at the University of Pennsylvania and practiced medicine in Philadelphia.

Mitchell had previous experience in chemistry, as he had experiences working in a chemical lab and a pharmaceutical store ran by another physician, Edward Parrish. At St. John's Lutheran College Mitchell became a chemistry professor and published research on chemistry and diseases [45]. As a physician, Mitchell practiced medicine in Frankford, Pennsylvania, just outside of Philadelphia. He was a prolific publisher and composed many works on the proper nature of disease, chemistry history, and medical pedagogy. Mitchell also contributed to medical literature by editing medical journals like *Western Medical Gazette*, as well as *Journal of Medical and Associate Sciences* until his death in 1865 [45]. In his chemical research, Mitchell historicized the teaching of chemistry and the American moment of success in chemistry [46], though last, not least, our own believed country claims a share in the need of praise, awarded to the cultivators of science." But he warned his students that non-professional health care providers, and even women, might have more mastery of chemistry than they did after graduating from university, but they should continue studying chemistry. Improving their chemical knowledge also improve their personal character and potentially make them wealthy, but also improved the fortunes of the nation [45–47].

The other cofounder of the Columbian Chemical Society was George F. Lehman who was a surgeon. Lehman was unique that he also had a medical doctorate from the University of Pennsylvania and also was a student of Benjamin Rush, where he eventually transitioned into medical practice as a physician. He worked a at the

Lazaretto, which was a special containment hospital where suffers from yellow fever were isolated and treated away from the general population of Philadelphia [48, 49]. He was the youngest physician working at the Lazaretto, starting his career there at age twenty-one and working there for the next twenty-five years. He also worked at the Quarantine Master of Philadelphia, and was praised for his work with contagion by the Council of Philadelphia, as well as the Board of Health, as he had successes treating many different types of fevers, in addition to small pox.

Lehman speculated in print about why corruption occurred, as well as how the airs produced disease, and why patients suffered from diseases related to urination. Lehman researched why the body produced stones. He dedicated his thesis at the University of Pennsylvania into studying biliary calculi and graduated in 1813 [50]. His dissertation was "An Inquiry into the Causes, Symptoms, and Cure of Biliary Calculi" where he examined ideas about the fluids of the body to understand why bodily stones, specifically in this case stones of the intestines, and if there was anything that could be done to help patients. He first started with a discussion about the liver as the organ where foods and other matter are transformed into the blood, which vitalized the body. The liver was important to running the chemistry of life [51], "this theory of the liver, which has been said to be only chimerical will be in a very short time generally adopted. It is simple and satisfactory." In this conception of bodily chemistry, the spleen, as well as the liver and gallbladder, worked to remove wasted fluids from the body.

Biliary stones had little literature about them, Lehman found, and many chemists and physicians wanted to write more about them, even lamenting this fact at the end of their careers [52]. Lehman, like many medico-chemists in this book, saw the body has heavily reliant on the idea of sympathy in defining disease, where one irritated part of the body influenced the other, an interconnect system [53, 54], "the human body is one great whole, so intimately connected that no particular part can act with regularity without consent of the others. Stimuli on impressions applied to one part producing motion, will excite it on others." Foods and other stimulating materials will affect the whole system and take that excitement to other parts of the body [53]. The liver could have problems as there could be too much excitement in the system, or an excess of something else, and it is not able to circulate vitality into the body. And once liver functionality was lost, it was hard to get back. A blocked or malfunctioning liver often led to the production of stones.

Bad bile, originating from the liver, made the rest of the body ill. Bad bile could cause problems in the stomach in the bile and blood ducts of the body, and these problems would also produce stones. Lehman, as students often did in medical dissertations, cited major authorities. He cited the work of George Cheyne

Shattuck,[5] a prominent physician in the Massachusetts medical community who had examined biliary calculi. Shattuck had studied biliary stones and found them large and bulky, which made them difficulty to circulate in the body. He had been able to examine these stones stuck in the body of a patient postmortem, who had found to have massive stones in his gallbladder [55].

Lehman himself had been able to examine passed gallstones and put them into chemical testing, as his friend donated some biliary stones to Lehman. He exposed the stones to heat, where they left a distinctive "wax" and could be destroyed simply by being placed between the fingers, giving Lehman the impression that these stones were just old bile [56]. Lehman knew that stones in the body were caused by all sorts of other explanations as well, such as imbibing alcohol, the constitutions of women over a certain age, episodes of emotions, sympathy problems, broken down circulation. At the age of fifty, Lehman theorized, their vital forces seemed to break down, especially in the power their bodies had to circulate their fluids and this fact, he claimed, was obvious to anyone, as people who were young seemed to not have to suffer from stones [56].

As a student educated in chemistry he brought in the latest chemical literature of his day to his dissertation. He quoted material from Antonine Francois Fourcroy, as well as William Sanders, who defined the bile [57], "It is of a yellowish green colour, of a bitter taste secreted from the blood by the liver." Physicians and chemists considered the nature and purpose of the bile and attempted to break the bile down chemically, like a chemist would decompose a substance to its elemental parts [58]. Thomas Thomson was also able to study the chemical nature of bile, and studied stones of the body as well. Many medico-chemists wanted to break down the bile, like they did urine, to understand how stones formed out of this bodily fluid.

Much like the work of Alexander Marcet and William Wollaston, Lehman tried to chemically categorize the types of biliary stones that he was able to collect. He said that there were six groupings of calculi found in the bile. The first group he called bilious hepatic. The second group was the heptic-adipocious. The third group was the cystic bilious. The fourth and fifth group were, respectively, the cortisal and the cystic adipocerous calculi. The final group was the mixed cystic. Lehman found that women suffered from cystic adipocerous at higher rates than men [58].

He speculated that chemistry held out hope for treating biliary stones. The remedies needed to be applied to the patient as the correct disease stage. In the first stages of the disease the patient needed to have a hot soak or strong medications that caused the body to expel the stone. If the body was producing painful spasms, Lehman, like Benjamin Rush, encouraged the patient to use the lancet to breathe the vein. But, unlike Rush, blood letting should be a remedy cautiously used as

[5] There needs to be more biographical information about this highly interesting figure. A good site for general information appears in his papers: https://www.masshist.org/collection-guides/view/fa0462.

it [59], “Bloodletting is of extreme importance, and should be resorted to immediately provided the pulse is tense.” The lancet would stop the irritation of the body and cause the body to calm itself in the intestines, so a stone did not form. Pain medications like opium were encouraged, as well as other medications that removed irritants from the body [59].

Lehman thought that there were curative medications that would be of use to the physician as well. He divided these cures into two groups, those interventions that were chemically rooted, and those that were mechanically rooted [60]. Mechanical interventions included physical acts that removed irritants from the body, like enemas, purgatives, opium, or blistering. Chemical interventions, though there were not many, were medications that could be used to adjust the system. Adjustments in the bodily system including acids that broke up the stone or changed the nature of the bile [61]. These chemical remedies for biliary stones resembled those medications proposed by chemists to treat urinary stones that were discussed in the previous chapter on Marcet.

Since Lehman served in a leadership role in the Columbian Chemical Society, we can see him contributing to the culture of chemical medicine in Philadelphia. His interest in bodily stones in his dissertation likely encapsulated his interest in chemical analysis of bodily diseases and stones and would likely be of interest to members in the society. The Columbian Chemical Society had an optimistic culture regarding chemical medicine. At the Columbian Chemical Society, members from the Chemical Society of Philadelphia continued to engage in chemical research and debate. Those members from the Chemical Society that continued over to the Columbian Chemical Society included important physicians in the Philadelphia medical community including chemist-physicians like Robert Hare and John Redman Coxe, as well as other physicians like Charles Caldwell.

Chemical societies like the Columbian Society ended in the first quarter of the nineteenth century, but new chemical societies would form out of community interests in chemistry in Philadelphia, such as the American Chemical Society that would emerge in the 1870s.[6] Historians like Miles argue that the War of 1812 distracted students and pulled them into the war effort. Other reasons for the ceasing of societies like the Columbian Society was the expenses of publishing their proceedings. The problem of graduating students also hurt the society, as many of its productive members returned home or left the city after graduation [62]. Historians have framed the importance of these chemical communities to shaping the medical culture of Philadelphia, while equating it with a type of chemical-medical masculinity, or as Wyndham Miles put it [63], “we do not know what led Mitchell and Lehman to organize a new society, but a logical assumption is that they received their inspiration from the early Chemical Society of Philadelphia. Several members of the new organization—Edward Cutbush, John Redman Coxe,

[6] See the American Chemical Society’s website for an interesting history: https://www.acs.org/about/history.html#:~:text=The%20First%20100%20Years,Chickering%20Hall%20in%20New%20York.

Adam Seybert, Charles Caldwell, John Syng Dorsey, Robert Hare, Henry Jackson and Robert Patterson—had belonged to the old society, and some of the young members were native Philadelphians who had undoubtedly watched the old society in the years when they had been growing into manhood." Many of the members were also part of other intellectual groups, such as the Philadelphia Medical Society, which demonstrates the interdisciplinary nature of medical chemistry in early nineteenth century Philadelphia.

7.2 Intellectual Life in Nineteenth Century South Carolina

Charleston was one of the largest cities in America during the late eighteenth and early nineteenth century, as it was a metropolitan city because of the freight and immigrants that were arriving in its port. Charleston was the third largest city in America, behind Boston and New York at the turn of the eighteenth century. In comparison, the historian Sally E. Hadden estimated that Charleston, in relation to other large cities in the 1790s, there were 1540 homes in Charleston by estimate, compared to 4600 in Philadelphia, and 2100 in Boston, and 3500 in New York [62].

Immigrants from France, England, and other places in Western Europe arrived in Charleston to flee political upheavals and revolutions. Planter and white elites also arrived in Charleston from the Caribbean seeking fortune and white supremacist culture, as Charleston was immersed in slavery. The port of Charleston and Charleston landing saw millions of imported slaves from Africa during the late eighteenth century [64, 65]. Because of its nature as a port city, Charleston became a hub for metropolitan ideas and diversity of thought [66]. One example of intellectual life in Charleston was the Literary and Philosophical Society. The society met in Charleston and discussed the latest scientific, medical, chemical, agricultural, and literature topics. Often their topics would focus in the promotion of slavery and white supremacy [67]. Many of the members were interested in chemistry, especially medical chemistry, and they promoted its potential, much like intellectual societies in Philadelphia. Members also discussed the problem of urinary stones in their own publications. Discussions of urinary stones represented another facet of boundary object, where the problem of urinary stones brought together intellectuals from different groups to discuss the problem [68].

The Literary and Philosophical Society of Charleston was made up of several contributing members that were physicians or established intellectuals in the city. A Huguenot Frenchman, John Linnaeus Edward Whitridge Shecut, who often shortened his name to JLEW in publications, and was the first president of the Literary and Philosophical Society. Shecut's family escaped religious persecution of protestants in France and eventually settled in South Carolina. He was born in Beaufort, South Carolina in 1770. His family were friends with the Ramsays, and Shecut received some of his earliest medical education under the famous physician David Ramsay. After an apprenticeship under David Ramsay, he received a medical doctorate at the University of Pennsylvania. Historians of South Carolina medicine, like Joseph Ioor Waring, argued that Shecut left the university without a degree because there are no records of him graduating [70].[7] Shecut did go on to publish many works on botany, yellow fever, medical chemistry, and medical topography [70].

In his 1819 work *Medical and Philosophical Essays*, Shecut discussed his ideas about medicine and how diseases spread [71]. In his *Essays*, he promoted the importance of science to the local and economic interest in science and he complained that Charleston's intellectual life could be much fuller [72]:

> It is a source of sincere regret to the lovers of science, that its progress has not been as rapid as the progress or topographical improvement of the city, with respect to its trade and commerce. For upwards of a century from its first settlement, science and literary may be said to have continued in a very languishing condition in Carolina. And although its history has recorded men of the first talents in the field and cabinet; and also several men of eminence in the science of medicine, there are but few moments of their literary labours, that have been preserved for succeeding ages.

But he did highlight the importance of the South Carolina Medical Society, which had started in 1789. He felt that the Medical Society made Charleston society and intellectual life much better [72]. Shecut highlighted the work of other South Carolina physicians who contributed to the intellectual culture of Charleston, such as John Drayton and David Ramsay [73–77].

The Literary and Philosophical Society was first led by a chemist and the society started to formally meet during 1809. The first leader of the society was Charles Dewar Simons. Simons was the first person at South Carolina College to teach chemistry. Simons was killed in an accident while crossing a swamp near Columbia and drowned. Simons horse might have fell into a ditch or he fell off his horse, but the horse was able to swim to safety but he did not. He had called to his "servant" (likely his slave) to save him, as the servant had gone to a house to try and warm himself prior to trying to cross the swamp. The conditions were icy and snowy. The servant, then referred to in the *South Carolina Gazette* as a "negro" tried to rescue Simons from the tree that he had braced himself to in the water, and Simons tried to swim to the servant, where they eventually both perished either by the coldness of the water or they both drowned. Unfortunately, both men perished at the Granby Swamp [78]. His chemical abilities were respected as the Trustees appointed him the first chair of chemistry at the College with a unanimous vote. But he only served the position for about a year, and Edward Darrell Smith was hired to replace Simons. Because of members like Simons, Shecut bragged that the society was respectable [78], "the Society, soon became highly respectable, with a large increase of members." The Society continued to raise its profile by having public lectures. Lecture topics included discussions of both natural histories, as well as demonstrations in philosophy (this usually included experiments), and the

[7] Waring was not able to find any record of Shecut being awarded a degree. See pages 303 and 371 in [69].

public came out to lectures in large numbers. Shecut promoted that Simons was likely hired by South Carolina College because of his successful public lectures [71, 79].

Another active chemist that participated in the society was Alexis De Carendeffez, who also published under the name Baron De Carendeffez. Like Felix Pascalis, Alexis De Carendeffez immigrated from San Domingo, but during the French Revolution. He made his way to Charleston after his arrival in New York sometime near 1800. De Carendeffez made his military mark in Yorktown, fighting with the French Army, which was at that time allied with the American army against the British. His heroism was commemorated in the historical work of another physician from Charleston, Alexander Garden [80]. Garden was not the well-known botanist and corresponded with the Royal Society with the same name, but a military leader who lived in St. Thomas Parish South Carolina. There are some records of people referring to him as Dr. but Waring noted that there were no explanations why he had that title as he was not a medical doctor [81]. Garden published an account of the American Revolution where he talked about the gallantry of Carendeffez [82]:

> …while seated on a barrel of powder, saw a shell from the enemy fall within two feet of his position. The Soldiers who were in the Battery, expecting immediate explosion, ran off in every direction. The intrepid youth remained unmoved. The expected catastrophe, however, did not follow—the fuse of the shell was, in its light, extinguished. This being perceived by the fugitives, the Battery was immediately reoccupied, when Captain Lemery, the commanding Officer, addressing himself to the youth, who still retrained his seat, said—'You young rogue, why did you not fly the impending danger? Why not embrace a change for life?' 'Because, Captain,' he heroically replied, 'my duty required that I should make a distribution of ammunition, and not desert my post, and fly like a poltroon!'

As a professional chemist, De Carendeffez published a lot of chemical research regarding paints and making paints [83]. There are also examples of his collaboration with the Chemical Society of Philadelphia, as he worked with James Wood House, who was leading the Chemical Society at that time, and the two men worked together to study the properties of water. Carendeffez shared an interest in the chemical compositions of the concretions of the body [84]. Writing prior to Alexander Marcet *Essay*, Carendeffez put concretions that he found logged in the lungs to chemical analysis. The patient was having trouble breathing because of their occupation as working in masonry, specifically the cutting of rocks. Carendeffez drew analogies between stones found in the lungs and those that were produced in the urinary organs. He read the paper writing up his work with the stones in 1802 in New York, as he was a member of another intellectual society, the Physical Society of New York, and it was later published in 1803 in *Medical Repository*. The patient had severe coughs, trouble taking in air, and passed blood, and had tremendous sweats in the evenings while trying to sleep [85]. The patient began expelling miniature stones, but as many as an estimated two hundred. He framed the expelling of stones in the patient that he held that [85], "a suspicion had arisen that these stony concretions were formed from the dust, inhaled while he was at working in shaping quarry-stones, by his mallet and chisel, for the purposes of architecture."

Similar to the example of Marcet and Astley Cooper working together to study stones, Carendeffez received pulmonary stones from another physician, Samuel Latham Mitchell. Mitchell was highly interested in medical chemistry and was active in publishing medical knowledge as he served as editor at *Medical Repository*. Mitchell taught chemistry at Columbia University in New York and frequently wrote letters with characters mentioned in this book including Felix Pascalis, Edward Darrell Smith, and Thomas Cooper [45, 86, 87]. The article contained chemical analysis of the stones he received from Mitchell that were expelled from the patient. After applying chemical reagents, like acids, he noticed that the stones responded by changing their color, and their overall shape changed as well. Like Marcet, he also put the stones to heat, and the stones were destroyed, though some shards transformed into producing a glass composition. Carendeffez proposed that any animal material in the stones had also been destroyed during his chemical analysis and they had lost some of their density. He also subjected the stones to a taste test, and found that after he pulverized them he described the taste as [88], "…a sour but agreeable taste." After performing his chemical research, he was confident that he found the correct chemical composition of the stones [88], "*…that these pulmonary stones or concretions are true PHOSPHATE OF LIME*."

After making the chemical identification of stones Carendeffez proposed that he could use chemical substances to stop such pulmonary stones from occurring. Chemists thought that the type of stones that Carendeffez found, the phosphate of lime stone, were thought to occur because of excess salt in the body, and these salts same from animal flesh or vegetables that patients consumed. The salt needed to be removed from the body and transported out of the body by its fluids. If there was not enough phosphate in the body, stones would form. Acids made of phosphorus removed stones from the body and kept salts equalized [89], "Hence, when there is no excess of phosphoric acid in the blood and secretions, we so often find concretions similar to these in the kidneys, in the bladder, in the bronchia, and in the lungs, and in other places." Because of his knowledge of the chemical literature, Carendeffez argued that the stones that were removed from the patient's lungs had similar chemical characteristics to other concretions that were produced in the body.

Carendeffez tried to excite the medico-chemical community to continue to investigate the stones of the body and emphasized the healing potential of acids to treat stones, such as acids like phosphoric, as well as oxalic substances. Philosophers had discovered by experiment that phosphoric, in addition to oxalic acid, presented stones from forming in the body [89]. Acid regents revealed to chemists the type of stone they were examining, as the work of Edward Darrell Smith, William Hyde Wollaston, and Alexander Marcet revealed how effective chemical testing was in identifying concretions of the body. Carendeffez promoted the success of chemical testing in the medico-chemical literature [89]:

> All these facts which I have seen and derived from my own experiences, in submitting these concretions to the action of different acids, and all the others which I have gathered from experiments made on calculi of the kidneys and bladder, convince me that most reliance is to be placed on the 'OXALIC and PHOSPHORIC ACIDS' for destroying these terrible concretions.

Patients, Carendeffez argued, should be prescribed either phosphoric or oxalic acid, despite their difficulty in consumption, as they often tasted poorly [89]. These acids were the most successful chemical remedy [89], "more stimulating upon the living parts, without having a proportional action upon the stones…I therefore recommend them to their notice and trial, as promising to do much in the cause of humanity, both in the form of drinks and injections."

During the turn of the nineteenth century, the Literary and Philosophical Society became more popular in Charleston. It attracted many members of the professions including religious clergy, plants, politicians, lawyers, doctors, and academics. By 1819, the Literary and Philosophical Society had grown to around 138 persons claiming membership [90]. The Society had expanded to have both a library and a museum, and these two institutions continue on in today. The intellectual culture of the Society branched out into objects, and those in membership, such as Carendeffez, circulated bodily concretion for testing, as demonstrated above in the summary of his article about pulmonary stones.

The Junior Literary Society was an expansion branch of the Literary and Philosophical Society for young people that emerged in 1814. This group did not last long and it ceased in 1819 [90]. Other attempts at engaging children in intellectual life failed, as David Ramsay tried to found the Ramsay Library Society, but this group also failed. It is unclear why intellectual societies for children failed in Charleston. Shecut tried to frame that these notable failures were not indicative of the intellectual abilities of the children of Charlestonians, but of their intense interest in learning [90]. For Shecut, the other point to having active intellectual societies like the Literary and Philosophical Society was key to promoting national power of the young United States. Having quality education and intellectual culture in Charleston should improve the prestige of the nation and [91],

> …prove to the world, that her soil is by no means unfavorable to the generation or cultivation of the arts and science, and that her sons want but the appropriate stimulus to their labours, that of public patronage, to shield them from loss, while endeavoring to raise her literary fame to a level with that of the most favoured nations, examples are yet to be seen in the most unpardonable apathy and shameful neglect of her citizens….Until the Carolinians are aroused to the formation of a permanent national character; and until the utility and vital importance of the arts, sciences and literature, form a predominant feature of that character, these things must and will remain, the reproach of South-Carolina.

The power of chemistry to national power was not a unique argument, as Thomas Thomson had also thought that the national power of Britain would be amplified by chemical education [92].

By 1814, the Literary and Philosophical Society met formally for the first time. They were led in 1814 by the well-respected botanist, and practicing physician,

Stephen Elliot. Elliot earned his medical degree in 1791 from Yale University and grew up in Beaufort, on the coast of South Carolina. He had established himself as a successful botanist and was awarded a medical doctorate from the Medical College of South Carolina in either 1824, or in 1825, where he also taught botany [93]. Elliot published a widely utilized guide to the uses of plants in South Carolina entitled *A Sketch of the Botany of South Carolina and Georgia* between 1816 through 1824, coming out one part at a time [93, 94]. Though Elliot died in 1830, another intellectual society was named in his honor that started in 1853. The Literary and Philosophical Society was ultimately reformed into the Elliot Society. While leading the Literary and Philosophical Society, he gave a speech promoting the usefulness of science, which included chemistry, and published this speech in 1814. Elliot promoted science as valuable to patriotism and economic interests and wanted his society to aspire to the prominence of intellectual societies in Europe. He promoted the importance of science in America, and he believed that men in his society could achieve great scientific discoveries [95, 96]:

> In Europe, where the pursuit of science has long been a cherished, and a fashionable occupation, and where the humbler of literary and scientific men has become so great as almost to crowd and jostle on the road, societies have been formed to promote the study of each distinct branch of knowledge. But with us it has been deemed advisable to unite in one Society all who should be willing to associate in our labors; while by arranging our members into different classes, and assigning to each class distinct and determinate objects, each individual will find himself co-operating with associates, having common views and occupation.

He made typologies of science and organized them from surgery to history [96]. Elliot saw the promotion of the many types of sciences that he laid out were a type of national strength that was crucial to knowledge production and integral to building the reputation of the United States.

Elliot saw chemistry as explanatory to diseases, such as the gasses in the air contributing to diseases like yellow fever. He went on to further argue that something in the airs contributed to disease and chemistry held the answer to ending contagion [97, 98],

> Chemistry ascertains the nature and properties of those airs or gases, which exist in the atmosphere, and perhaps pervade all nature, it analyses the composition of the atmosphere and endeavours to elucidate its changes. Hence those modifications of the air, which constitute the science of meteorology, the result of combinations of the gaseous fluids, varied probably by electric and magnetic influence, become objects of chemical enquiry.

In addition to chemistry, intellectuals in the group also used the society to further arguments that supported slavery and white supremacy. One publication that came from a member of the society included the work of J. H. Guenebault. Guenebault published a translation of the Natural History of the Negro, written by the French philosopher Julien-Joseph Virey [98–100]. The Literary and Philosophical Society of Charleston was society that was generally in support of slavery and white supremacy.

In Charleston, Elliot, Carendeffez, Guenebault, and other members wanted to transform the intellectual culture of the city and encourage discussions of medico-chemistry, among other scientific topics. Members like Carendeffez completed chemical research on bodily concretions. In the excitement about science, members also promoted national strength from the building of scientific culture. These discussions also included the promotion of slavery and white supremacy [99–101]. The Literary and Philosophical Society pursued the prestige associated with intellectual societies in Europe and wanted to create scientific knowledge in wider avenues of public engagement, such as lectures and publication. Intertwined in that cultural production was chemical research of the bodily concretion and the study of the body's fluids in explaining disease.

7.3 Medical, Chemical, and Surgical Knowledge Production in London: The Medico-Chirurgical Society

Chemists, surgeons, and physicians in London came together at the turn of the nineteenth century to build a society where they could discuss the latest ideas that were important to science and disease. Alexander Marcet, Astley Cooper, William Hyde Wollaston, and other characters mentioned in this book helped to found the Medico-Chirurgical Society in Holborn, London that started around 1805. Membership even included American members like Benjamin Rush [102]. Ultimately, the Medico-Chirurgical Society would transition into the Royal Society of Medicine later in the middle of the nineteenth century [102]. The Medico-Chirurgical society was a meeting place for all of the healing disciplines. However, the society was not without problems. For instance, the first leader of the Medico-Chirurgical Society would not step down from office, and ruled the society for around twenty years. The society met and papers were read by its members, and those papers were later published in *The Medico-Chirurgical Transactions*, which started in 1809 [102]. The meetings of the Medico-Chirurgical Society brought together practitioners from London hospitals and other practicing surgeons and physicians. Chemists and other intellectuals interested in science also attended the events.[8] There was also a library and eventually object collecting began [102]. Many of the members founded medical schools and other teaching organizations

[8] The Medico-Chirurgical Society was created prior to the founding of the Chemical Society of London in 1841. Please see the historical section on the Royal Society of Chemistry's webpage https://www.rsc.org/about-us/our-history/ as well as Brock WH (2013) The London Chemical Society 1824. Ambix 8:133–139. During the early nineteenth century the Worshipful Society of Apothecaries of London would have been active and in the same community, see Hunting P (2004) The Worshipful Society of Apothecaries of London. Postgrad. Med. J. 80:41–44.

around London as well [102]. By 1834, the Medico-Chirurgical society was recognized by King William IV and became the Royal Society of Medicine. The Royal Society of Medicine is a major organization in regulation and training physicians in London today [102]. Many of the members of the Medico-Chirurgical society were also members, or worked with members, of the Royal College of Surgeons [102]. The liminal space between the two organization facilitated discussions regarding bodily stones and medical chemistry between chemists, physicians, and surgeons in London. The liminal space was reflected through the publications of the two groups as well. Members of both organizations worked on blockages found in the body, like stones, as well as problems in the body's fluids which produced disease.

John Yelloly, a member of the Medico-Chirurgical society was a leader in the organization, but also interested in researching the stones of the body and promoting medico-chemistry. John Yelloly had an interesting life, as uncle, Nathaniel Davidson, served as his parent after both Yelloly's parents had passed away. Yelloly was originally from Northumberland and was born in 1774. He studied medicine at the University of Edinburgh and either graduates in 1796, or possibly 1799. After his medical training he practiced in London, in the Aldersgate neighborhood. Yelloly was active in medical institutions in London including the London Hospital, as well as the Medical Society of London, and contributed to training doctors in the city and working on improving pharmacy. Many of Yelloly's friends were members of the Medico-Chirurgical Society, as he was friends with Alexander Marcet, as well as Astley Cooper, as they had been members of other intellectual societies in London including the Edinburgh Club, as well as John Hunter's Pow Wow Club. More research needs to be conducted about both clubs, especially the nature of the Pow Wow Club. The two societies were meant for physicians and other healthcare providers [103, 104]. Yelloly was a close friend of Marcet, as he revived a share of his papers and books when Marcet died. Yelloly was visited periodically by other surgeons and physicians, like Astley Cooper, Thomas Thomson, as well as Willian Sanders [104, 105].

Bodily stones were discussed at meetings of the Medico-Chirurgical Society, as they appeared later as read papers in the *Transactions* [105]. The paper that Yelloly read about a stone that was manually removed from a female patient resembled many of the cases of stones that are in this book and it continued the idea that women would eventually no longer need the operation for the stone. The Medico-Chirurgical Society had meetings where health care providers also brought specimens to the meetings, such as the reading of the 1815 paper about the women who suffered from a stone; Yelloly brought the stone that was removed from the patient. The stone was a boundary object, that received interests from the many different attenders at the meeting. This was likely not the only paper that involved a bit of show and tell from the presenter.[9]

[9] More research needs to be completed about how many papers involved objects.

Yelloly's paper was a report that had been communicated to him by letter by a surgeon. The collaboration between physicians and surgeons at the Medico-Chirurgical Society as they frequently communicated information to each other. Yelloly communicated the case on Mr. Hopke's account, who was a surgeon. The patient, only named as "J. M." had problems urinating throughout her life, starting around age seven, and expressed blood in her urine after jumping [106]. The problems continued into her twenties and but when she was married the problems seemed to start, and then re-started themselves when she found out she was pregnant. The blood in the urine continued and the urine had a strong odor. There was speculation this might be from a venereal disease, like gonorrhea, but the physician only speculated [106]. The pregnancy seemed to proceed without issue and her body continued to have incontinence [107].

But the pain continued and physicians treated that pain with narcotics, but the amount of incontinence also decreased, but problems with urination continued and [106], "always [had] a considerable deposit of a purulent appearance in the urine, as was observed likewise to have been the case in her third pregnancy." J. M. expelled calculi in her body [107], "She now passed, with temporary relief, several substances [calculi] of more than half an inch in length and described as resembling the fur from the inside of a tea-kettle." The surgeon acted after the patient continued to suffer from stones and used a probe, which Hopke put into the patient's urethra and removed a stone. Yelloly described the characteristics of the stone as being large [107], "... "eighteen grains...rather flat, and was somewhat of an oval form, having the longest diameter rather more than six-eights of an inch in length, and the shortest about five-eights." Yelloly continued to recount Hopke's action in search for stone in J. M.'s body [107], "He found by means of a probe induced into the bladder, and also by the finger introduced into the vagina, that there was a very large stone in the bladder." Though the patients felt better after the removal of stones, the break from pain was short lived.

The patient discussed her pain with Hopke, and she thought that she might pass another stone. Hopke discussed the state of the patient [108], "...per vaginam, he found the calculus sticking in the urethra, but there was no opening between the urethra and vagina; and remarked at the time, that if there had been such an opening, he would have felt himself warranted in the enlarging it, for the purpose of removing the stone." Another surgeon from London attended the patient named Mr. Headington, and the stone was extracted manually by one of the surgeon's hands [108]. The patient reported feeling better after the removal of the stone but continued to have incontinence. The patient experienced the same type of pain when she had another child. Unfortunately, after the healthy birth of the child, the woman became ill, perhaps a cold, her body slowed its production of milk, and she died on Christmas day in 1813. J. M.'s unfortunate death came three weeks after the birth of the new child.

The stones from the case were put to chemical analysis. The stones were described in the paper in order to assign them a grouping or category being [109], "...of an irregular surface, and of a flattened oval shape, having two little rounded projections at the extremity by which it passed from the urethra." The weight of the stone was estimated to be around 3 oz and Yelloly further described its chemical makeup [109], "it is composed principally of uric acid, disposed in close concentric lamellae, having no perceptible nucleus; and a considerable portion of its surface is covered with a mixture of phosphate of lime, and ammoniaco-magnesian phosphate."

Additional chemical studies were published in Yelloly, including those published in other institutions periodicals. Yelloly published a case regarding concretions in the Royal Society, an institution that focused mostly on science and natural philosophy, in their *Philosophical Transactions*. The Royal Society was a higher status intellectual institution in London that had Royal patronage since Charles II, that featured leaders in its history including Robert Boyle, Humphrey Davy, and Isaac Newton [111]. The case he shared involved a woman suffering from a humongous stone and he argued that it was difficult to manually extract these especially large stones [110, 111]. In his paper he thought that manually extracting stones that were bulky could injure the patient [111–113], "...the parietes of the urethra, in all such cases, are injured by ulceration, which by making an opening from this tube into the vagina, facilitates the exit of the calculus." He commented that the patient's bladder was vulnerable to damage, as well as creating a dangerous hole in the middle of the vagina, as well as the urethra [112, 113]. Yelloly's position that manual extraction was something that would be of value to female patients, and allow them to eventually avoid a lithotomy, furthered Marcet and Astley Cooper's views about women [113, 114]:

> when we consider the shortness of the female urethra, the thickness of its parietes, the want of resistance from contiguous parts, and the facility with which it has in many instances been distended by mechanical means*, it is the less to be wondered, that the long continued pressure of a calculus in the female has often produced a sufficient dilation for its removal.

He continued to make comparisons, this time evoking cases of stones that were extracted from the biliary system [113, 114], "we have the great dilatability of the biliary ducts evinced by the magnitude of the concretions which have passed through them into the duodenum." The point Yelloly continued to make is that women, one day, would be able to not require a lithotomy.

Like Marcet, Yelloly studied statistically the rates of stones in hospital populations and the number of bodily stones made up the rolls of admitted patients. In his papers that were published in the publications of the Royal Society, he studied patients and stones at the Norwhich Hospital, and both of these papers were read to the Royal Society [114, 115]. Yelloly was very interested in the same questions that Marcet would later inquire about in his *Essay*. He made more wide-ranging generalizations than Marcet, and Yelloly proposed that the prevalence of stones in both England Wales is about one person to 188,000 people. But the Norwhich hospital seems to have a higher rate of suffers of the stone, or he estimated that there were two cases of stones in around twenty-one thousand people. He also compared that to one cases of stones in about thirty-eight thousand people in the city of London. Yelloly did not speculate on why the numbers were different in each area [116, 117], "I regret that but little advances have been made, in a knowledge of the circumstances on which a tendency to calculus complaint depends; and I am not aware of such differences of air, water, soil, or habits of life having yet been detected, as can justify us in attributing the prevalence of stone, in the Norfolk district, to any of those causes." Humoral theory was in the background of Yelloly's thinking, as he attributed possible differences in the suffering of stones to the patient's constitution. He tried to explain the prevalence of stones in regards to the constitution [117], "A constitutional predisposition to the occurrence of calculus diseases unquestionably exists in certain families." Yelloly referred to the work of William Prout, a medico-chemist, who was discussed earlier in this book. Prout had found that suffering from urinary stones often ran in families, which spoke to constitutional issues that ran in families [117]. Yelloly hoped that by keeping records of what people suffered from urinary stones, and what populations suffered from stones, doctors could understand the problem better. Yelloly had the same aims as Marcet in keeping better records, or any records at all, regarding the numbers of suffers of concretions. Like Marcet, Yelloly hoped to spur record keeping of diseases of the concretions [118]:

> If I might venture, however, to make the suggestion, I would respectably submit, how subservient our public hospitals, the boasts and ornaments of the countrymen might be made to important statistical inquiries, buy a more extended system of registry, than is at present usually adopted, either in the metropolis, or in the country; and how conductive to pathological improvement, the information would be, which they might be so readily enable to furnish.

Yelloly wanted to create a record keeping system of the number of stone suffers. He hoped that by better record keeping other diseases that contributed the formation of body concretions could be determined by doctors. More information could illuminate underlying causes that contributed to stone production [119],

> In the instances which I have mentioned, it would therefore appear, that the tendency to produce calculous complaints, is greater in towns than in the country; and if this should prove to be the cases generally, it would seem to indicate the existence, in children more particularly, of a connection between some diathesis which prevails in towns, (probably the scrophulous) and the tendency to the secretion or deposition of lithic acid, on which the origin of the urinary calculi so much depends. I have not had it in my power to ascertain, whether the greater disposition of towns to calculous complaints, applies more extensively than I have mentioned.

Yelloly not only continued Marcet's desire to keep better record of stone sufferers, but also continued to put urinary stones to chemical analysis.

Yelloly put the stones that he had access to from the Norwhich hospital into groups by their chemical makeup. The stones mostly resembled the chemical groupings of William Hyde Wollaston and Alexander Marcet, but Yelloly believed that his chemical testing of the stones might be more sophisticated than the work that Marcet's had been done previously [120],

> I found, however, that none of the calculi contained in it [Marcet's *Essay*] were divided, and that the experiments instituted by our laments colleague (of which an account was published in his work on Calculous Diseases) were therefore necessarily confined to the outer surface, except in cases where the calculus had been broken in the extraction, and its interior structure thus allowed to be seen.

Yelloly had a lot of stones to test, as he around three hundred and thirty stones. He was conversant in the chemical literature of other medico-chemists who had done research into the characteristics of urinary stones. He found a previously unknown chemical in the stones that he tested that he named "silex" [120, 121]. He believed that this new chemical that was discovered in the stones held a significant insight into understanding urinary stones. Yelloly was excited in promoting its importance [122], "And as if the love of exciting surprise and admiration by the marvelous, were not confined to the traveler, there is sometimes found in patients, however singular the fact may appear, an inclination to impose on their professional attendants, by the description or exhibition of something strange and anomalous." Discovering the chemical substance would save both patients and practitioners from mistakes regarding other false explanations of why bodily stones formed who could be [122], "...deceived, by the mistakes of patients, or their friends, in matters of an unusual nature." Like Marcet, Edward Darrell Smith, and other practitioners discussed in this book, Yelloly believed that information about urine and stone composition needed to be widely discussed and published so that medico-chemists could understand the primary chemicals in stone formation in the body [122].

Yelloly promoted the new stone classification he called silex [122]. The urinary stone that he based his new classification on was gathered from a child many years ago. Because of its characteristic in regards to color, Yelloly speculated at first that it might have been a type of oxalate urinary stone. When Yelloly subjected it to chemical testing he had to be very careful, as the stone was miniature, and only one round of chemical analysis could be performed at the chemical bench. While performing chemical research, Yelloly worked with another chemist, Michael Faraday [123]. Faraday was friends with Marcet, Yelloly, John Bostock, and many other members of the Medico-Chirurgical Society. He learned chemistry through the work of Alexander Marcet's wife, Jane Marcet, as discussed earlier in the book. Faraday received urinary stones from Yelloly and Yelloly helped Faraday enter into the membership of the Royal Society. The chemical findings of these stones were published in *Philosophical Transactions of the Royal Society*. Yelloly and Faraday had their chemical research findings recorded in the article. The heat they subjected the urinary stone that they named silex was destroyed by heat and the character of the stone changed, as some of it blew up when exposed to the heat. The rest of the stone turned into a type of silica [122].

Towards the conclusion of the paper, Yelloly quoted the work of a Mr. Copeland Hutchinson. Copeland and agreed with the constitutional effect of stones formation and perhaps an underlying cause of stones could be the disease of scrofula, a type of Tuberculous in today's medicine [124]. At the end of his article studying the nature of urinary stones among sailors, Copeland framed the importance of Yelloly,

Marcet, and many other medico-chemists in trying to determine the rates in which people suffered from stones [125],

> The statistical inquiries on the subject of urinary calculi which have been published in this country within in the last fourteen years, by Drs. Marcet and Prout, my Messrs. Richard Smith, Martineua, Crichton, Liston, and lastly, by Dr. Yelloly, cannot but be highly useful to the future inquirer and throw light upon the nature of this extraordinary disorder. I feel assured, indeed, with my lamented friend, Dr. Marcet, that it is chiefly in this way that the true pathology of the disease can ever be obtained, and consequently, the most efficacious mode of treatment.

The members of the Medico-Chirurgical Society were interested in studying how stones formed, performing chemical research, and understanding what people suffered from stones. They wanted to build networks and collections of stones in order to test stones and further refine chemical groupings. Members also wanted to understanding the frequency of stones in order to speculate if humoral ideas like a patient's constitution, or other lifestyle factors, contributed to disease. Doctors and chemists also wanted more information in regards to the appearance of stones to determine if any other diseases contributed to their formation. The Society itself was a venue for the discussion of stones, as well as their publication of papers in the *The Medico-Chirurgical Transactions*. Collaborations between physicians, chemists, and surgeons occurred in and outside of the society in order to build information and collections of urinary stones, and further chemical knowledge. Discussions of urinary stones, as well as other bodily stones, continued in London beyond disciplinary boundaries at the Royal College of Surgeons.

7.4 The Royal College of Surgeons, Collection Building, and the Need for Chemists

Many of the members of the Medico-Chirurgical Society also participated in the activities of the Royal College of Surgeons, also in London. There were friendships like Marcet and Astley Cooper, and other friends who seemed to appear in each other's papers. The line between surgery and medicine was very permeable because of the good relationships between these two societies. One person who worked between the two groups was William Clift. Clift was an artist, surgeon, and one of the main organizers of the extensive museum collection at the Royal College of Surgeons. The collection would ultimately form the basis of the new Natural History Museum in the nineteenth century [126]. One of the problems that began to occur in the early nineteenth century at the Royal College of Surgeons was an abundance of medical specimens, including urinary stones. The collection was getting so large it was hard to organize, but the members of the College wanted to promote and spread the extensive of the collection to other interested intellectual societies. The promotion and collaboration in managing and expanding the collection involved members of the Medico-Chirurgical Society, like Peter Roget. Roget was a chemist, physician, and intellectually curious person that participated in knowledge creation in many different disciplines. Clift and Roget corresponded about the collection at the Royal College of Surgeons [126–129],

> I am instructed by the President & Council of the Medical + [sic] Chirurgical Society to send you the accompanying preparation of a monstrous fetus, which we transmitted to them by Professor Maumoir [sic] of Geneva, of which an account is just published in the seventh Volume of the Medico-Chirurgical Transactions, + [sic] which they have the honor to present, in the name of the Society, to the Royal College of Surgeons,
>
> I have the honor to be
>
> Sir, your most obedient servant
>
> P. [Peter] M. [Mark] Roget, (Secretary).

Objects often traveled between the groups, much like the friendship and collaborations. The two institutions represented the best examples of boundary objects that this book explores. Clift was helpful at communicating information about objects and specimens, as his talent as an artist allowed him to draw many of the discussed objects at the meetings of the Medico-Chirurgical Society and his artworks were published in the *Transactions of the Medico-Chirurgical Society*. Clift worked with other members at the Medico-Chirurgical Society and got access to the society's library.

The Royal College of Surgeon has a fascinating history and its history a microstudy of the professionalization of surgery from craft knowledge [130]. The Royal College of Surgeons came into existence because of the need of surgeons who were versed in anatomy and physiology to gather, continue their education, regulate their profession, and train students.[10] Surgeons wanted to imitate the professional movement of physicians in the UK, and sought royal patronage to further their professional goals [130–132]. The College of Surgeons became the Royal College of Surgeons around 1800, through it only had regulatory power in London. It was not until 1843 that the institution became the Royal College of England, where it had more power, through the medical landscape of regulation was still unclear and there was lack of comprehensive professional oversight [132–134]. John and William Hunter were the champions of the Royal College of Surgeons, as their professional celebrity and achievements in anatomy bolstered support for the institution, and they provided many of the early anatomy examples and objects that formed the basis of the collection. The Royal College of Surgery eventually called its teaching anatomy and surgical examples the Hunterian [135].

Clift acted as type of agent who traveled between the Royal College of Surgeons and the Medico-Chirurgical Society. He was a fellow of both institutions [102, 136]. Clift, born in 1775, was originally from Cornwall and aspired to become a surgeon. He apprenticed himself to John Hunter and he paid Hunter through providing his drawings. Clift had an immense talent for art and he likely provided

[10] There are unfortunately not a lot of history of the Royal College of Surgeons. See the institutions own interpretation of its history: https://www.rcseng.ac.uk/about-the-rcs/history-of-the-rcs/.

the images in Marcet's *Essay*. The apprenticeship went well and Clift was released after a six-year training period [136–139].

After John Hunter died he left a roughly twenty-thousand-piece collection that included everything from anatomical models and objects, as well as urinary stones, were ultimately purchased by Parliament and turned over to the College of Surgeons in 1793 [136–139]. Clift was placed as the custodian of the collection as he helped negotiate the sale between John Hunter's widow, and family members, as well as Parliament. Clift also organized the first archival collection, as many of Hunter's personal papers were also transferred to the College of Surgeons.

He was a passionate leader of the Hunterian Museum. He left detailed inventory of the collection, especially as it was expanding.[11] Cliff was also an amazing artist and many of his sketches and images appeared in periodic literature. His artwork was not just limited to medical images, but he also published images in paleontology, geology, as well as everyday life scenes [139, 140]. He worked with many of the characters in this book in London, including John Bostock. Clift's life was memorialized in many leading medical and scientific publications of the day such as *The Lancet*, as well as the *Philosophical Transactions*.

Because Clift had such detailed records at the Hunterian, historians know a lot about the early history of the collection. Many types of exotic animals were added to the collection, including the 1806 purchase of tusks from elephants. Other additions included other animal concretions to the collection. Like [140, 141], "Seven various concretions ^v[sic] from the Bladders and stomachs of Quadrupeds." Clift's descriptions continued to list additional, distinctive [141] "concretions." He continued to make additional notions on the published description as marginalia [141]:

> Several of the Concretions were broken into Lamina or Concentric Shells, and all the Sections and pieces were counted singly. These have been since mended (1821) and sections brought together, which were in different lots, and two of them were rounded flint-pebbles which had some appearance of calculi. One other was a bad Cast of a Calculus in plaster of Paris. By this means, the number has been considerably reduced to 30 Boxes.

As the lead of the Hunterian, Clift was not only adding materials, but removing objects that no longer fit the collection, or deaccessioning. This was a difficult job, as the Royal College was often overwhelmed with submissions of distinctive medical objects that members had either collected or removed during surgery, and this was in addition to donations from other groups, like the Medico-Chirurgical Society. There are several notions and entries from Clift for objects that were donated to the collections, like bodily stones. In 1807, Everard Home, a famous surgeon, who later would be accused of plagiarizing John Hunter's work, donated a concretion and Clift listed it [141–144], College "A box containing 307 [...] Calculi, from a man age 77. Which weighted 9 oz. 7 1/2 drachms; and also Seven

[11] The Royal College of Surgery was an amazing research trip, and I acquired a lot of information about the history of the institution from an incredibly helpful archival staff. Also in regards to the citation for Cope's history: This citation was found fully completed, as reviews and other electronic editions omitted the publisher. The information was found in Loudon I (2000) Why Are (Male Surgeons) Still Addressed as Mr? BMJ 321: 1589–1591.

small ones which passed before alkalies, were used." The College became the reception point of the material objects that were discussed in intellectual societies across London.

The Royal College of Surgeons had to call on chemists to try and sort out all the many concretions they had in their expansive collections by the 1830s. William Thomas Brande was an active chemist at another intellectual society in London, the Royal Society [145]. Brande applied his chemical expertise and collaborated with another surgeon, William Blizard, to bring order to many of the concretions that Blizard donated to the Royal College of Surgeons. Blizard was an accomplished surgeon who was a member of the Royal College of Surgeons, as well as instrumental in establishing the Medico-Chirurgical Society [146]. There are specific examples of collaborations that can be found in the manuscripts of the Royal College of Surgeons. Listed in the collections of the Hunterian is concretion that is listed as "stone 16," which is also listed as 306 as it was part of the Hunterian. The concretion includes written contextual information [147], "An oblong intire [sic] calculus (about 2 oz 'in weight). 'From M^{rs}. Bliss' Presented by Sir W^{m}. Blizard 1811. Selected for analysis Aug. 29. 1833." A handwritten note indicates that it contained an "oxalate[sic] nucleus." Clift was able to organize many of the concretions into the existing chemical groups.

Historians like NG Coley has chronicled Clift's membership and work in the chemical community in London, as he was a member of the animal chemistry club [104]. He did fine concretions that did not fit common chemical grouping, and as this book as pointed out in earlier chapters, chemical groups of concretions were flexible and still being revised by the medico-chemical community. An example of a stone that did not fit existing chemical assemblies of concretions was "stone 14" which became "stone 170" as named in the Hunterian. Stones that were presented to the College often were accompanied by a chemical description were sometimes incorrect [147], "An undivided Calculus of 'Oxalate of Lime.' Not so. (Selected for Analysis Aug 29. 1833.)" The testing chemists often left a notion of their name beside the stones that they tested, or suspected were labeled incorrectly. Brande left his monograms as "WFB" besides concretions he was active in researching or thought were mislabeled.

The results of the chemical testing of the stones of the Hunterian ended up in the yearly report that Clift gave to the board that oversaw the Hunterian collection. Clift stated his findings [148]:

> the Chemical Analysis and descriptive Catalogue of the Hunterian Collection of Calculi and Concretions of the Human Subject have been so nearly brought to a completion as to admit of a general Comparison and Classification of these productions; and that the following Scheme has been adopted as the basis of this Arrangement:__[.]

The Royal College believed in chemistry as a way of knowing and improving their collection, and chemistry improved surgical knowledge as these same collections would be used to train aspiring surgeons [148, 149]. The emphasis on chemical analysis by surgeons was another example of concretions being boundary objects, as they brought practitioners were diverse fields together to solve a problem: what was the nature of stones that the Royal College had in their possession, and for chemists, how representative were their chemical taxonomies.

Clift, who had an interest in chemistry and was plugged into a community of medico-chemists interested in studying concretion, including Alexander Marcet, as well as William Prout proposed that there were eight chemical groupings of the stones of the collection [104]. The eight groupings were Phosphate of Life, Ammonia, and a combination of the two. Other groupings including the relatively new Xanthic Oxide Stone. Oxlate groupings included those of Lime, Magnesia, and Ammonia. Finally, there were stones that fit into the Cystic Oxide grouping [147]. Clift tried to organize both bodily concretions and the concretions of the Hunterian based on this group and reported this plan to the board in 1839 [147],

> And that certain alterations of the Cabinets destined for the reception of the Calculi, with trays of an uniform pattern and appropriate sizes have been ordered, and are in progress, so as the better to adapt them for the display and arrangement of the Calculi, corresponding to the preceding Classification; and that the ~~Board~~ [sic] Committee entertain the expectation that this part of the Catalogue will be in the Press, and the arrangement of the Calculi completed before the next Meeting of the Trustees.

Clift circulated letters and requests to continue to build the collection of stones for both professionals and aspiring surgeons. In 1816, Clift wrote a letter to another surgeon, Mr. Jeksinson, who was working a local hospital in London, and Clift communicated information about the location and formation of concretions in the lungs after the postmortem of "Tera Poo", whose lungs had been extremely painful to the patient in life [150], "...almost universal enlargement and induration of the lymphatic glands, particularly those of the Mesentery." Clift described the concretions that were removed from the lungs [150], "The natural structure of the Lungs had been almost entirely destroyed the substance was filled with Tubercles which varied... from the Size of a Pea to that of a Walnut, and many of these had proceeded to a State of Suppuration."

The Royal College would ultimately publish the results of not only their chemical testing, but a full report of their complete collections for public and professional consumption. The Hunterian collection continued to grow as well. After Clift, the collection was overseen by the famous future paleontologist and scholar Richard Owen. Under Owen's leadership, who was also a surgeon like Clift, the collection continued to expand [151]. The collection of stones was both extensive and important enough for the College to issue *A Descriptive and Illustrated Catalogue of the Calculi and Other Animal Concretions Contained in the Museum of the Royal College of Surgeons in London*, which contained several illustrated plates of concretions of the collection. These catalogues were published in the 1830s, prior to Clift's retirement in 1842, the same year that Owen took over managing the Hunterian [151, 152]. The work was massive as it was three volumes in length [152, 153]. The Royal College of Surgeons in Edinburgh, also published information about their collections as well, which also listed their collections of stones [154, 155].

An early catalogue was organized by Clift, but it was not exclusively organized by chemical categories but by the areas of the body that they commonly occurred in, or as Clift described as [156] "Diseases of the Urinary Organs." Clift continued to encourage the inclusion of chemical information in the published categories. In Volume I of *A Descriptive and Illustrated Catalogue*, there was contextual and historical information of the urinary stones that made up the Hunterian collection. Volume II contained descriptions of the extensive collection of animal stones at the College. Volume III, which was the final volume focused on the large collection of human stones that the College had collected over its history, as well as those it acquired from members of other institutions. The published *A Descriptive and Illustrated Catalogue* went into publication while the College's Hunterian collection was undergoing the transition of leadership between Clift and Owen [157]. While Clift remained at the Hunterian, he worked to get chemical information included in the published volumes, including the chemical research that had been completed on the stones that William Blizard donated. After 1842, Owen was in charge of the collection and museum at the Royal College of Surgeons, and participated in the activities of the Medico-Chirurgical Society, where he was elected in honorarium, a member [158]. Owen would go on to build an extensive publication repertoire based off the holdings of the Royal College of Surgeons [159]. These publications included his work on bodily stones, especially from animals like crawfish.

The first volume attempted to explain the overarching organization concepts of concretions. The preface listed all the organization groups of the concretions [160], "…first, an excessive, or an altered and vitiated secretion; and secondly, substances introduced with the food, and retained in the different parts of the alimentary canal. To the former class belong, not only the products of the true glandular organs, as the liver and kidney, but also those of the mucous and synovial surfaces." The preface went on to describe other categories [161], "namely those which are to be traced in the different parts of the alimentary canal." In regards to the two classification groupings, in the first, the most common stone was made up of chalk, which was often associated with gout patients. The second grouping was mostly made up of stones that had formed around objects that had entered the body somehow.

Volume I included information as to why there was a lack of chemical analysis performed in the stones from the early years of the collection, most of which were from the surgeon John Hunter. The volume contained the explanation, as well as a gentle historicization of chemistry [162], "owing to the imperfect state of chemical knowledge at that period, Mr. Hunter attempted no further arrangement of these bodies, than by simply referring them to the different organs from which they were taken." The chemical nature of the stones would gradually take on a larger, and more important role as the years went on, and Clift encouraged chemical research.

The Hunterian collection continued to grow in regards to concretion beyond the initial donations of Blizard and John Hunter. The chemical community added to the collection, as there were donations from chemists like William Brande. The Royal College wanted to build a complete collection of every stone imaginable, as this was one of the ideas that surgeons like Astley Cooper and chemists like Alexander Marcet thought would led to knowledge about why stones formed in the body, or perhaps better treatments for stones. The College was working in the shadow of the European Enlightenment and building complete collections were thought to be the useful form of solving problems [163]. *A Descriptive and Illustrated Catalogue* contained an enthusiasm that [161–164], "...have rendered the Collection of Human Urinary Calculi nearly complete."

Other intellectuals published studies about the collection of concretions at the Royal College of Surgeons. Brande was able to compile all of his chemical research from the Hunterian collection that he undertook while working with Clift and have it appear in the *Philosophical Transactions* in 1808. The concretion collection was researched again by another chemist, Thomas Taylor, during 1834–1838 [164–166]. Taylor was a chemist, and during 1845, he became a member of the Medico-Chirurgical Society, and he also was a member of the Royal College of Surgeons. He taught chemistry at the Middlesex Hospital, in addition to teaching chemistry to medical students [167, 168]. Taylor's chemical expertise were credited as giving order to an increasingly chaotic collection of bodily stones at the Hunterian [168, 169], "Within the last four years the entire Collection has undergone a complete revision, and the calculi have, for the first time, been arranged in a systematic order. Every specimen has undergone individual examination, as far as that could be done without injury to the calculus. The accomplishment of this undertaking was confided to Mr. Thomas Taylor, a member of the College, whose fitness for the task is proved in the manner of its execution; and the Council have much gratification in acknowledging the value of his services." Taylor's name is listed as author on most of the library index of *A Descriptive and Illustrated Catalogue*, though he is not given formal credit inside of the catalogues. According to *Plarr's Lives of the Fellows*, Taylor was paid to work on the project, with his fees being around one hundred pounds [167]. Taylor was trained as a surgeon and he performed chemical analysis for members of other intellectual institutions in London, such as the Royal Society.

The Board of the College was agnostic about including continuing to submit the collection of stones to the newest chemical analysis but decided to memorize their choices in the printed catalogue. The Catalogue memorized that [169, 170], "the composition of urinary calculi from the human subject has been of late years so thoroughly investigated, that it was scarcely to be expected that much additional information would be gained by the examination of that part of the Collection." Though the memorization was skeptical of new knowledge being produced by chemists like Taylor, there were new findings about the collection. One finding was the increase of chemical grouping to total a of nine. The collection was reorganized to reflect the new findings from the chemical testing in the 1830s, and Clift's scheme of stones that was put forth in his 1839 work was the organizing structure that was published in *A Descriptive and Illustrated Catalogue*. The College's collection of stones strove to be comprehensive, as it was an Enlightenment creation, and it strove to have every type of stone [170]. Because of chemical testing, chemist were able to show the chemical groups where the College had no specimens, and their deficits are listed in the catalogue [169], "Carbonate of Lime. Of this species of calculus the Museum poses no specimen." There were a lot of

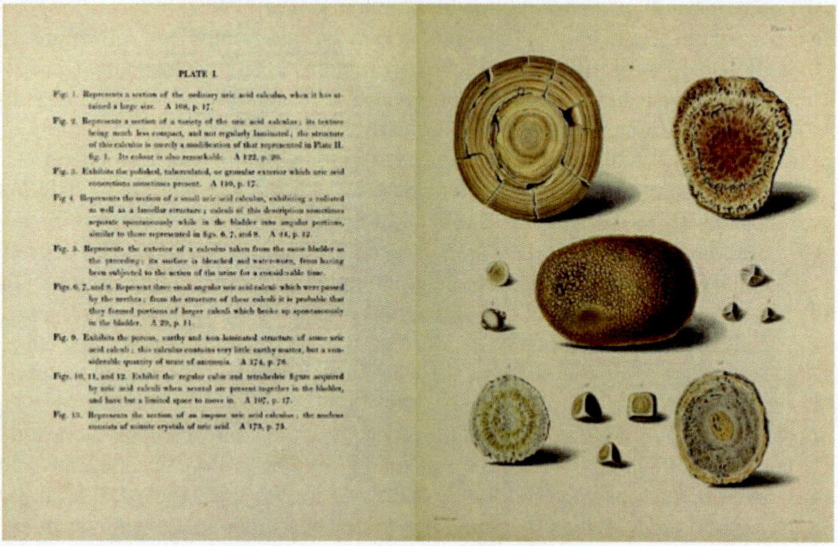

Fig. 7.2 Plates from *A Descriptive and Illustrated Catalogue* showing different types of stones, with explanatory labeling of the plate

rich images in *A Descriptive and Illustrated Catalogue* of the stones in the collection. Some example plates and description from the first volume of *A Descriptive and Illustrated Catalogue* in Figs. 7.2, 7.3 and 7.4. The plates provided a vivid narrative of the collection, and these plates could also be used to educate surgeons, physicians, and chemists about the typologies of stones.

After the publication of the first volume in 1842, the remaining two volumes were published at the same time, and were released in a single publication. The focus of Volume II is the concretions of animals, from mammals, to reptiles, as well as birds. There were concretions from whales and primates as well [171]. Clift was interested in collecting concretions from animals and he donated his objects to the College [171], "P 5. Dried masses of the semifluid urine of the Boa Constrictor. This substance consists principally of suburate of ammonia. *Presented by W. Clift, Esq.*" The amount of concretions became overwhelming to the College, as donations picked up, and it was difficult to differentiate between animal and human stones. The catalogue denoted examples that were thought to be human, but were found to be animal in origin [172], "P8. This calculous was placed by Mr. Hunter among the human urinary concretions, but as it only differs in size from the two last described specimens, it is most probable that it has also a similar origin." Other examples include a stone group, the "Urate of Potass" that was thought to have human origins, but was actually from a reptile [172], "The following calculi are the only instances that have been hitherto discovered or uric acid in combination with potass. They bear a striking resemblance to each other in their

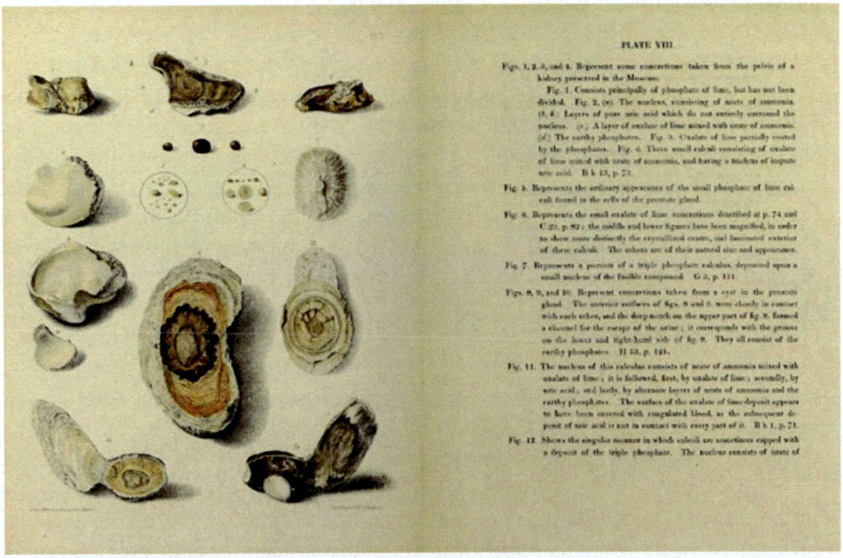

Fig. 7.3 Another plate from *A Descriptive and Illustrated Catalogue* discussing the different types of stones extracted after surgical action

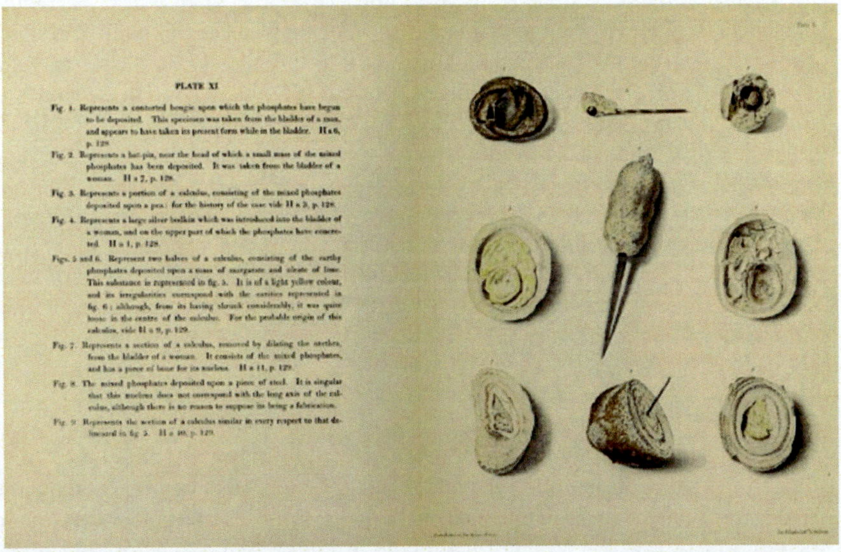

Fig. 7.4 Stones removed from bladders during surgical operations; plate was included from *A Descriptive and Illustrated Catalogue*

structure and general appearance, and have been doubtless taken from the urinary bladder of some of the species of Iguana that are found in South America."

In the final third of the *A Descriptive and Illustrated Catalogue* the very rare stones of the human body are described and thoroughly examined. Their chemical groupings were half the size of those of the first volume in describing urinary stones. Citing the work of the medico-chemists Thomas Thomson, the catalogue of human stones from the stomach, mouth, intestinal tract, lungs, joints, veins, and other bodily ducts, organized these stones into five chemical classes. The Catalogue cautioned that only careful medico-chemists could spot the differences between these stones, and other stones discussed in the work, such as urinary stones. The older stones in the collection from Blizard often confused chemists and it took a lot of careful descriptive work [173], "two impure cholesterine calculi, taken from the same gallbladder. One is oblong and flattened at either end; the other is angular." Careful chemistry revealed not only the true nature of specimens in the collection, but helped to properly identify what stone the patient was suffering from (with enough care from the practitioner).

A Descriptive and Illustrated Catalogue concludes with a final section about stones that did not any of the chemical taxonomies, or could not be properly placed (or unknown). Some of the most difficult stones to identify and research are those that originated in the lungs, as well as the eyes. *A Descriptive and Illustrated Catalogue* listed another deficit in their collection, and they knew that the stones existed because there were accounts of the stones appearing in studies of bodily concretions, such as the lung stone from Wollaston that Marcet analyzed, or accounts of the stones of the eye. An ophthalmologist, William Mackenzie, published a work on diseases of the eye, which recounted case histories of patients suffering from eye calculi [174]. He analogized this case of calculi in the eye to those stones that formed in the salivary glands [175]. The College did have a concretion in their collection that originated from a woman's lungs [176], "B1. A stone spit out of a woman's lungs. From Dr. Grew's Collection—*Sloanian Ms. Catagloue*. Phosphate and carbonate of lime. *British Museum*." The last volume ends with a listing of all the gout stones, which the catalogue credits as originally written about and analyzed by William Hyde Wollaston. The collection is missing other types of stones, especially those found in the body's veins, and the College only has four stones from gout sufferers [177].

7.5 Institutions of Stones

Stones, including those of humans and animals, were a boundary object that united some practitioners from intellectual institutions from Charleston in the southern part of America, to the northern metropole of Philadelphia, to across the Atlantic in London. Organizing, naming, memorializing the various stones of the body united the interests of physicians, chemists, and surgeons, as well as other intellectuals in order to promote the power of chemistry. Even surgeons, who are often portrayed as not having as much interest in chemistry as physicians, saw its power, but remained as skeptical as most chemically minded physicians about the limits of chemistry in curing bodily stones. Stone collecting continued the Enlightenment project of acquiring one of everything in order to understand the natural world, of making diseases, as the historian Peter Dear described the project of science, "intelligible" [178]. On the British Museum's website, on their webpage discussing their Enlightenment origins, that state that collecting spirit, though myopic, involved the untold stories of enslavement, and Eurocentrism convey the ways of knowing were collecting, "While Enlightenment thinking and collections provided the foundations for much of our present understanding of the history of human cultural achievement, they also tended to tell that story from a predominantly European perspective."[12]

Though the story of turn of the nineteenth century is by no means the story of urinary stones, these objects did play a role, among other specimens and diseases, in intellectual institutions collecting knowledge and trying to improve the treatment of diseases by better understanding. Urinary stones, and bodily concretions in general, show the permeability of the boundaries that intellectuals draw around disciplines and questions, and show how having plastic boundaries improves scientific and medical research. Societies, such as the Medico-Chirurgical Society and the Royal College of Surgeons, as well as their publications, institutionalized research and working that was occurring informally in friendships and collaborations going on between doctors, surgeons, and chemists. The optimism around the power of chemistry to construct better classifications, improve life, and cure diseases was of interest to institutions like the Literary and Philosophical Society of Charleston, as well as the Philadelphia based Chemical Society and the Columbian Society, promoted chemistry to members who came from many different disciplines. Planters, medical students, and philosophers heard discussions of human concretions, among other objects, as they served witness to the excitement around the potential for chemistry to improve human lives. It was the publications of the intellectual societies discussed in this chapter, along with the wide-ranging question of what, if anything can be done with chemistry to improve our knowledge of bodily stones, that made intellectual societies on both sides of the Atlantic accessible to even remote physicians, like Edward Darrell Smith living in Western South Carolina, so that he could read enough chemical and medical literature to become well-read enough to teach chemistry [178–180].

References

1. Chaplin JE (1993) An Anxious pursuit: agricultural innovation and modernity in the Lower South, 1730–18156. University of North Carolina Press, Chapel Hill
2. Schiebinger LL (2004) Plants and Empire: colonial bioprospecting in the Atlantic World. Harvard University Press, Cambridge

[12] See https://www.britishmuseum.org/collection/galleries/enlightenment.

3. Delbourgo J (2006) A most amazing scene of wonders: electricity and enlightenment in early America. Harvard University Press, Cambridge
4. Safier N (2008) Measuring the new world: enlightenment science and South America. University of Chicago Press, Chicago
5. Parish SS (2006) American curiosity: cultures of natural history in the colonial British World. University of North Carolina Press, Chapel Hill
6. Delbourgo J, Dew N (eds) (2007) Science and Empire in the Atlantic World. Taylor & Francis, New York
7. Gänger S (2021) A Singular Remedy: Cinchona across the Atlantic World, 1751–1820. Cambridge University Press, Cambridge
8. Miles W (1950) Early American Chemical Societies: 1. The 1789 Chemical Society of Philadelphia 2. The Chemical Society of Philadelphia. Chymia 3:95–113
9. Penington J (1790) Chemical and economical essays. Joseph James, Philadelphia
10. Williams WD, Miles W (1990) Old chemistries: John Pennington's 'chemical and economic essays.' Bull Hist Chem 8:18–22
11. Bell WJ Jr (1957) John Redman, Medical preceptor, 1722–1808. Pa Mag Hist Biogr 81:157–169
12. Bates RS (1958) Scientific societies in the United States. Columbia University Press, New York
13. Gibson SS (1982) Scientific societies and exchange: a facet of the history of scientific communication. J Libr Hist 17:144–163
14. Carson J (1869) A history of the Medical Department of the University of Pennsylvania. Lindsay and Blakiston, Philadelphia
15. Kilbride D (1999) Southern medical students in Philadelphia, 1800–1861: science and sociability in the "Republic of Medicine." J South Hist 65:697–732
16. Woodard C (2011) America: a history of the Eleven Rival Regional cultures of North America. Penguin Books, New York
17. Smith EF (1918) James Woodhouse: a pioneer in chemistry, 1770–1809, John C. Winston Company, Philadelphia
18. Calloway CG (2015) The victory with no name: The Native American Defeat of the First American Army. Oxford University Press, New York
19. Edelstein SM (1959) The chemical revolution in American from the pages of the "medical repository." Chymia 5:155–179
20. Smith EF (1923) Early scientists of Philadelphia. Pa Mag Hist Biogr 47:1–27
21. Jackson J (2005) A world on fire: a Heretic, an Aristocrat, and the race to discovery oxygen. Viking, New York
22. Johnson S (2008) The invention of air: a story of science, faith, revolution, and the Birth of America. Riverhead Books, New York
23. Toulmin SE (1957) Crucial experiments: Priestley and Lavoisier. JHI 18:205–220
24. Philadelphia Gazette, June 21, 1797
25. Kelly HA, Burrage WL (1920) Cyclopedia of American biography, vol 1. The Norman Remington Company, Baltimore, p 894
26. Dun JA (2016) Dangerous neighbors: making the Haitian Revolution in Early American. University of Pennsylvania Press, Philadelphia
27. Manzano-Aguilera JM (2008) Slavery and medicine in the Caribbean at the end of the *Ancien Règime*. Soc Hist 33:383–401
28. Furstenberg F (2015) When the United States spoke French: five refugees who shaped a Nation. Penguin Books, New York
29. Pascalis F (1807) Account of an abscess of the liver terminating favorably by the evacuation through the lungs. Med Repos 17:500–503
30. Rey T (2017) Dr. Pascalis and the Making of American medicine. In: The Priest and the Prophetess: Abbè Ouvière, Romaine Rivière, and the Revolutionary Atlantic World. Oxford University Press, New York

31. Waring JI (1964) A history of medicine in South Carolina: 1670–1825. South Carolina Medical Association, Charleston, p 349
32. Pascalis-Ouviere F (1802) Annual oration, delivered before the Chemical Society of Philadelphia, January 31 1801. John Bioren, Philadelphia
33. Pascalis-Ouviere F (1802) Annual oration, delivered before the Chemical Society of Philadelphia, January 31 1801. John Bioren, Philadelphia, p 38–39
34. Pascalis-Ouviere F (1802) Annual oration, delivered before the Chemical Society of Philadelphia, January 31st, 1801. John Bioren, Philadelphia, p 39
35. Driggers EA (2015) Boundary stones: morbid concretions and the Chemistry of Early Nineteenth Century Medicine. Ph.D. Dissertation, University of South Carolina: https://scholarcommons.sc.edu/etd/3646/
36. Pascalis-Ouviere F (1802) Annual oration, delivered before the Chemical Society of Philadelphia, January 31, 1801. John Bioren, Philadelphia, p 47
37. Pascalis F (1818) Desultory remarks on the cause and nature of the black colour in the human species; occasioned by the case of a white women suddenly turned black. Med Repos 4:366–371. The original citation was found in the reference below
38. Barnes LL (2005) Needles, herbs, gods, and ghosts: China, healing, and the West to 1848. Harvard University Press, Cambridge
39. Pascalis F (1818) Desultory remarks on the cause and nature of the black colour in the human species; occasioned by the case of a white women suddenly turned black. Med Repos 4:366
40. Pascalis F (1818) Desultory remarks on the cause and nature of the black colour in the human species; occasioned by the case of a white women suddenly turned black. Med Repos 4:367
41. Pascalis F (1818) Desultory remarks on the cause and nature of the black colour in the human species; occasioned by the case of a white women suddenly turned black. Med Repos 4:368
42. Pascalis F (1818) Desultory remarks on the cause and nature of the black colour in the human species; occasioned by the case of a white women suddenly turned black. Med Repos 4:371
43. Memoirs of the Columbian Chemical Society of Philadelphia (1813). Isaac Pierce, Philadelphia
44. Miles WD (1975) The Columbian Chemical Society Hist Stud Phys Sci 5:145–154
45. Kelly HA, Burrage WL (1920) Cyclopedia of American biography. The Norman Remington Company, Baltimore 1:805–806
46. Thomas Jefferson University Archives and Special Collections, MS 060, Thomas D. Mitchell Papers (1837–1862)
47. Jensen WB (1989) Thomas Duché Mitchell and the Chemistry of Principles. Bull Hist Chem 5:42–48
48. Postal Guide, Philadelphia Post-Office: Containing Postal Information by the United States Post Office (1900). Howard March, Philadelphia. Lehman's biography begins on page 9 of the book
49. Barnes DS (2006) The Great Stink of Paris and the Nineteenth-Century Struggle Against Filthy and Germs
50. Lehman GF (1813) An inquiry into the causes, symptoms and cure of Biliary Calculi. Medical Dissertation, University of Pennsylvania. University of Pennsylvania Rare Books Room, Philadelphia
51. Lehman GF (1813) An inquiry into the causes, symptoms and cure of Biliary Calculi. Medical Dissertation, University of Pennsylvania. University of Pennsylvania Rare Books Room, Philadelphia, p 3
52. Lehman GF (1813) An inquiry into the causes, symptoms and cure of Biliary Calculi. Medical Dissertation, University of Pennsylvania. University of Pennsylvania Rare Books Room, Philadelphia, p 5
53. Lehman GF (1813) An inquiry into the causes, symptoms and cure of Biliary Calculi. Medical Dissertation, University of Pennsylvania. University of Pennsylvania Rare Books Room, Philadelphia, p 6
54. Driggers EA (2021) "The voice of the stomach": the mind, hypochdriasis and theories of dyspepsia in the nineteenth century. Hist of Pysc 32:58–99

55. Lehman GF (1813) An inquiry into the causes, symptoms and cure of Biliary Calculi. Medical Dissertation, University of Pennsylvania. University of Pennsylvania Rare Books Room, Philadelphia, p 8
56. Lehman GF (1813) An inquiry into the causes, symptoms and cure of Biliary Calculi. Medical Dissertation, University of Pennsylvania. University of Pennsylvania Rare Books Room, Philadelphia, p 9
57. Lehman GF (1813) An inquiry into the causes, symptoms and cure of Biliary Calculi. Medical Dissertation, University of Pennsylvania. University of Pennsylvania Rare Books Room, Philadelphia, pp 13–14
58. Lehman GF (1813) An inquiry into the causes, symptoms and cure of Biliary Calculi. Medical Dissertation, University of Pennsylvania. University of Pennsylvania Rare Books Room, Philadelphia, p 23–24
59. Lehman GF (1813) An inquiry into the causes, symptoms and cure of Biliary Calculi. Medical Dissertation, University of Pennsylvania. University of Pennsylvania Rare Books Room, Philadelphia, p 28
60. Lehman GF (1813) An inquiry into the causes, symptoms and cure of Biliary Calculi. Medical Dissertation, University of Pennsylvania. University of Pennsylvania Rare Books Room, Philadelphia, p 29
61. Lehman GF (1813) An inquiry into the causes, symptoms and cure of Biliary Calculi. Medical Dissertation, University of Pennsylvania. University of Pennsylvania Rare Books Room, Philadelphia, pp 30–34
62. Miles WD (1975) The Columbian Chemical Society. Hist Stud Phys Sci 5:146
63. Miles WD (1975) The Columbian Chemical Society. Hist Stud Phys Sci 5:145–154
64. Hadden SE (2011) The business of justice: merchants in the Charleston Chamber of Commerce and Arbitration in the 1780s and 1790s in Wells JD, Green JR The Southern Middle Class in the Long Nineteenth Century. Louisiana State University Press, Baton Rouge
65. Quintana RA (2018) Making a Slave State: political developments in Early South Carolina. University of North Carolina Press, Chapel Hill
66. Berlin I (2000) Many thousands gone: the first two Centuries of Slavery in North America. Harvard University Press, Cambridge
67. O'Brien M, Moltke-Hansen D (eds) (1986) Intellectual life in Antebellum Charleston. University of Tennessee Press, Knoxville
68. Driggers EA (2019) The chemistry of blackness: Benjamin Rush, Thomas Jefferson, Everard Home, and the project of defining blackness through chemical explanations. Crit Philos Race 7:372–391
69. Bowker GC, Timmermans S, Clarke AE, Balka E (eds) (2015) Boundary objects and beyond: working with Leigh Star. MIT Press, Cambridge
70. Waring IJ (1964) A history of medicine in South Carolina: 1670–1825. South Carolina Medical Association, Charleston, p 303 and 371
71. Looney JJ (2009) Papers of Thomas Jefferson: retirement series, vol 6. University of Virginia, Charlottesville, pp 179–180
72. Shecut JLEW (1819) Medical and philosophical essays. A. E. Miller, Charleston, p 39
73. Shecut JLEW (1819) Medical and philosophical essays. A. E. Miller, Charleston, p 43
74. Drayton J (1802) A View of South Carolina, As respects her natural and civil concerns. W. P. Young, Charleston
75. Friedman LJ, Shaffer AH (1979) History, politics, and health in early American thought: the case of David Ramsay. J Amer Stud 13:37–56
76. Stephens LD (2000) Science, race, and religion in the American South: John Bachman and the Charleston circle of naturalists, 1815–1895. University of North Carolina Press, Chapel Hill
77. Faust DG (1977) A sacred circle: the Dilemma of the intellectual in the Old South, 1840–1860. Johns Hopkins Press, Baltimore
78. Labrode M (1859) History of South Carolina College: from its incorporation December 19, 1801, To November 25.1857. Peter B. Glass, Columbia, pp 59–63

79. Shecut JLEW (1819) Medical and philosophical essays. A. E. Miller, Charleston, pp 47–48
80. Stephens LD (2003) The literary and philosophical society of South Carolina: a forum for intellectual progress in Antebellum Charleston. SC Hist Mag 104:154–175
81. Garden A (1822) Anecdotes of the revolutionary war in America. A. E. Miller, Charleston
82. Waring IJ (1964) A history of medicine in South Carolina: 1670–1825. South Carolina Medical Association, Charleston, pp 335. Carendeffez has a biography on pp 184–185
83. Garden A (1822) Anecdotes of the revolutionary war in America. A. E. Miller, Charleston, p 207
84. Carendeffez A (1807) An account of a Rouge or red paint for the use of the toilet. Med Repos 10:304–306
85. Carendeffez A (1803) An analytical description of certain Stony Concretions. Med Repos 6:258–261
86. Carendeffez A (1803) An Analytical Description of Certain Stony Concretions. Med Repos 6:259
87. Smith EF (1922) Samuel Latham Mitchill—A Father in American Chemistry. Journ of Indus and Engineer Chem 14:556–568
88. Carendeffez A (1803) An analytical description of certain Stony Concretions. Med Repos 6:260
89. Carendeffez A (1803) An analytical description of certain Stony Concretions. Med Repos 6:261
90. South Carolina Historical Society. Literary and Philosophical Society of Charleston. February 8, 1832 (43/527) South Carolina Historical Society, Charleston
91. Shecut JLEW (1819) Medical and philosophical essays. A. E. Miller, Charleston, pp 54–55
92. Cohen IB (1976) Science and the growth of the American Republic. Rev Pol 38:359–398
93. Waring IJ (1964) A history of medicine in South Carolina: 1670–1825. South Carolina Medical Association, Charleston, p 211
94. Elliot S (1816–1824) A sketch of Botany in South Carolina and Georgia. J. R. Schenck, Charleston
95. Sanders AE, Anderson WD Jr (1999) Natural history investigations In South Carolina from colonial times to the present. University of South Carolina Press, Columbia
96. Elliot S (1814) An address to the literary and Philosophical Society of South Carolina; Delivered in Charleston, On Wednesday, the 10 August, 1814. W. P. Young, Charleston, pp 3–4
97. Elliot S (1814) An address to the literary and Philosophical Society of South Carolina; Delivered in Charleston, On Wednesday, the 10 August, 1814. W. P. Young, Charleston, p 16
98. Elliot S (1814) An address to the literary and Philosophical Society of South Carolina; Delivered in Charleston, On Wednesday, the 10 August, 1814. W. P. Young, Charleston, p 5
99. Driggers EA (2019) The chemistry of blackness: benjamin rush, thomas jefferson, everard home, and the project of defining blackness through chemical explanations. Crit Phil Race 7:372–391
100. Virey JJ (1837) Natural history of the Negro Race. Trans Guenebault JH. D. J. Dowling, Charleston. Special Collections, University of South Carolina, Columbia
101. Faust DG (1979) A Southern Stewardship: the intellectual and Proslavery argument. Amer Qrtly 31:63–80
102. Hunting P (2002) The history of the royal society of medicine. The Royal Society of Medicine Press, London, pp 1–67
103. Cope Z (1959) The Royal College of Surgeons England: A History. London: Anthony Blond.
104. Coley NG (1967) The Animal Chemistry Club; Assistant Society to the Royal Society. Notes and Rec 22:173–185
105. Richardson G (2002) A Norfolk Network Within the Royal Society. Notes and Rec of the Roy Soc of Lon 56:27–39
106. Yelloly J (1815) Particulars of a Case of Which A Large Calculus Was Removed from the Urethra of a Female Without Operation; With Examples of Analogous Cases. Med-Chir Trans 6:574–676

107. Yelloly J (1815) Particulars of a Case of Which A Large Calculus Was Removed from the Urethra of a Female Without Operation; With Examples of Analogous Cases. Med-Chir Trans 6:575–576
108. Yelloly J (1815) Particulars of a Case of Which A Large Calculus Was Removed from the Urethra of a Female Without Operation; With Examples of Analogous Cases. Med-Chir Trans 6:576
109. Yelloly J (1815) Particulars of a Case of Which A Large Calculus Was Removed from the Urethra of a Female Without Operation; With Examples of Analogous Cases. Med-Chir Trans 6:576–577
110. Yelloly J (1815) Particulars of a Case of Which A Large Calculus Was Removed from the Urethra of a Female Without Operation; With Examples of Analogous Cases. Med-Chir Trans 6:578
111. Tinniswood A (2019) The Royal Society: And the Invention of Modern Science. Basic Books, New York
112. Tefekli A, Cezayirli F (2013) The History of Urinary Stones. In Parallel with Civilization Sci W J Nov: https://doi.org/10.1155/2013/423964
113. Yelloly J (1815) Particulars of a case of which a large calculus was removed from the urethra of a female without operation; with examples of analogous cases. Med-Chir Trans 6:581
114. Yelloly J (1815) Particulars of a case of which a large calculus was removed from the urethra of a female without operation; with examples of analogous cases. Med-Chir Trans 6:582
115. Yelloly J (1829) Remarks on the tendency to calculous disease: with observations on the nature of urinary concretions, and an analysis of a large part of the collection belonging to the Norfolk and Norwich Hospital. Phil Trans Roy Soc 119:55–81
116. Yelloly J (1830) Sequel to a paper on the tendency to calculous diseases, and on the concretions to which such diseases give rise. Phil Trans Roy Soc 120:415–425
117. Yelloly J (1829) Remarks on the tendency to calculous disease: with observations on the nature of urinary concretions, and an analysis of a large part of the collection belonging to the Norfolk and Norwich Hospital. Phil Trans Roy Soc 119:64
118. Yelloly J (1829) Remarks on the tendency to calculous disease: with observations on the nature of urinary concretions, and an analysis of a large part of the collection belonging to the Norfolk and Norwich Hospital. Phil Trans Roy Soc 119:67
119. Yelloly J (1829) Remarks on the tendency to calculous disease: with observations on the nature of urinary concretions, and an analysis of a large part of the collection belonging to the Norfolk and Norwich Hospital. Phil Trans Roy Soc 119:66
120. Yelloly J (1829) Remarks on the tendency to calculous disease: with observations on the nature of urinary concretions, and an analysis of a large part of the collection belonging to the Norfolk and Norwich Hospital. Phil Trans Roy Soc 119:69
121. Yelloly J (1829) Remarks on the tendency to calculous disease: with observations on the nature of urinary concretions, and an analysis of a large part of the collection belonging to the Norfolk and Norwich Hospital. Phil Trans Roy Soc 119:79
122. Yelloly J (1830) Sequel to a paper on the tendency to calculous diseases, and on the concretions to which such diseases give rise. Phil Trans Roy Soc 120:419
123. Thomas JM (1991) Michael Faraday and The Royal Institution: the genius of man and place. Taylor & Francis, New York
124. Moulis G, Martin-Blondel G (2012) Scofula, The King's Evil. Cana Med Assc J 12:1061
125. Hutchinson AC (1831) A further inquiry into the comparative infrequency of calculus disease among sea-faring people, with some observations on their frequency in Scotland. Medico-Chir Trans 16:94–127
126. Hussey K (2014) The Hidden Museums of the Royal College of Surgeons of England. H-Net accessed 11/7/2022: https://networks.h-net.org/node/9782/discussions/20346/hidden-museums-royal-college-surgeons-england
127. Pearce JMS (2020) Roget and his Thesaurus. Hektoen Intern accessed 11/7/2022: https://hekint.org/2020/09/29/roget-and-his-thesaurus/

128. Dobson J (1955) William Clift, F. R. S.: first conservator of the Hunterian Museum. Roy Soc Med 48:323–325
129. Roget P to Clift W (1816) Museum Letters at Royal College of Surgeons, London. 1816 RCS-MUS/5/6/6, letter 128
130. Brown M (2023) Emotions and surgery in Britain, 1793–1912. Cambridge University Press, Cambridge
131. Pelling M, with White F (2003) Medical conflicts in early modern London: patronage, physicians, and irregular practitioners, 1550–1640. Clarendon Press, Oxford
132. Roberts MJD (2009) The politics of professionalization: MPs, medical men, and the 1858 Medical Act. Med Hist 53:37–56
133. Loudon I (2007) The royal college of surgeons. J Soci Med 100:397
134. Payne T, Toms JD, Zaidi A, Thrumurthy SG (2021) The history of surgery and surgical training in the UK. Pak J Med Sci 37:1532–1535
135. Pearce N (ed) (2015) William Hunter's world: the art and science of eighteenth-century collecting. Routledge, New York
136. (1848) Medico-Chirurgical Transactions 31
137. Dobson J (1954) William Clift. Heinemann Medical Books, London
138. Conservators of the Hunterian Museum. Royal College of Surgeons, London
139. Dictionary of National Biography
140. Meli DB (2015) The rise of pathological illustrations. Bull His Med 89:209–242
141. Clift W (Undated) Rough copy of purchases of specimens of natural history & by the College for the Museum. Viewed at the Royal College of Surgeons, London
142. Tricker J (2012) Saints and Sinners: Everard Home. R Col Surg Eng Bull 94:316–318
143. (1802–1816) Presents to the Museum, from the Year 1802 to 1816; inclusive, now called donations, p 35. Viewed at the Royal College of Surgeons, London
144. Clift W (1833) Specimens of Calculi 1833. MS Viewed at the Royal College of Surgeons, London. Viewed at the Royal College of Surgeons, London
145. Spires CH (1969) William Thomas Brande, Leather expert. Ann Sci 25:179–201
146. McConaghey RMS (1958) Sir William Blizzard and his poems. Med Hist 2:292–297
147. Clift W. Specimens of Calculi 1833. Royal College of Surgeons MS 0007/1/6/1/18. Viewed at the Royal College of Surgeons, London
148. Clift W (Undated) Specimens of Calculi 1833. Royal College of Surgeons MS 0007/1/6/1/21. Viewed at the Royal College of Surgeons, London
149. Pickstone JV (2000) Ways of knowing: a new history of science, technology and medicine. Manchester University Press, Manchester
150. Clift W to Jenkinson (1816) William Clift to Mr. Jenkinson. Royal College of Surgeons, Museum Letters 1816 RCS-MUS/5/6/6. Viewed at the Royal College of Surgeons, London
151. Negus V (1966) History of the Trustees of the Hunterian Collection. E. & S. Livingstone LTD., Edinburgh and London, pp 32–36
152. Rupke NA (1994) Richard Owen: biology without Darwin. University of Chicago Press, Chicago
153. (1842–1845) A Descriptive and Illustrated Catalogue of the Calculi and Other Animal Concretions Contained in the Museum of the Royal College of Surgeons in London. Viewed at Royal College of Surgeons. RCS-MUS/7/19/1
154. (1949) William Clift, F. R. S. (1775–1849). Nature 163:946
155. (1836) Catalogue of the Museum of the Royal College of Surgeons of Edinburgh. Neill and Company, Edinburgh
156. Clift W (Undated) Catalogue of the contents of the Museum of the Royal College of Surgeons in London. RCS-MUS/7/9/1. Viewed at the Royal College of Surgeons, London
157. Cope Z (1959) The Royal College of Surgeons of England. Anthony Blond, London. See footnote.
158. (1850) Medico-Chirurgical Transactions 33

159. (1845) Descriptive and Illustrated Catalogue of the Physiological Series of Comparative Anatomy Contained in the Museum of the Royal College of Surgeons in London, Volume III Part II: Connective and Tegumentary Systems and Peculiarities. RCS-MUS/7/11/7, pp 303. Viewed at the Royal College of Surgeons, London
160. (1842) A Descriptive and Illustrated Catalogue of the Calculi and Other Animal Concretions Contained in the Museum of the Royal College of Surgeons in London. Viewed at Royal College of Surgeons. RCS-MUS/7/19/1, preface
161. (1842) A Descriptive and Illustrated Catalogue of the Calculi and Other Animal Concretions Contained in the Museum of the Royal College of Surgeons in London. Viewed at Royal College of Surgeons. RCS-MUS/7/19/1, pp iii–iv
162. (1842) A Descriptive and Illustrated Catalogue of the Calculi and Other Animal Concretions Contained in the Museum of the Royal College of Surgeons in London. Viewed at Royal College of Surgeons. RCS-MUS/7/19/1, p v
163. MacGregor A (2007) Curiosity and enlightenment: collectors and collections in sixteenth to the nineteenth century. Yale University Press, New Haven
164. Cooper A (2007) Inventing the Indigenous: local knowledge and natural history in early modern Europe. Cambridge University Press, Cambridge
165. Brande W (1808) A letter on the differences in the structure of Calculi. Phil Trans 98:223–243
166. Home E (1808) Some observations on Mr. Brande's paper on Calculi. Phil Trans 98:244–248
167. Plarr's Lives of the Fellows. RCS: E003216. Accessed 11/9/2022: https://livesonline.rcseng.ac.uk/client/en_GB/lives/search/detailnonmodal?qu=%22rcs%3A+E003216%22&d=ent%3A%2F%2FSD_ASSET%2F0%2FSD_ASSET%3A375399%7E%7E0&ic=true&rt=false%7C%7C%7CIDENTIFIER%7C%7C%7CResource+Identifier&h=0
168. Taylor T (1845) On some new species of animal concretions. Lon, Edbin, Dubl Phil Mag J Sci 28:36–46
169. (1842) A Descriptive and Illustrated Catalogue of the Calculi and Other Animal Concretions. Contained in the Museum of the Royal College of Surgeons in London. Viewed at Royal College of Surgeons. RCS-MUS/7/19/1, p vi
170. Wilson P (1995) An enlightenment science? Surgery and the royal society. Clio Med 29:360–386
171. (1845) A Descriptive and Illustrated Catalogue of the Calculi and Other Animal Concretions Contained in the Museum of the Royal College of Surgeons in London. Viewed at Royal College of Surgeons. RCS-MUS/7/19/1, Volume 2, p 133, 135–136
172. (1845) A Descriptive and Illustrated Catalogue of the Calculi and Other Animal Concretions Contained in the Museum of the Royal College of Surgeons in London Viewed at Royal College of Surgeons. RCS-MUS/7/19/1, vol 2, p 141
173. (1845) A Descriptive and Illustrated Catalogue of the Calculi and Other Animal Concretions Contained in the Museum of the Royal College of Surgeons in London [Part II Calculi from the Urinary Organs of the Lower Animals and Part III Concretions found in Other Parts of the Body. RCS-MUS/7/19/4. Viewed at the Royal College of Surgeons, London, p 171
174. Mackenzie W (1830) A practical treatise on the diseases of the eye. Longman, Rees, Orme, Brown, & Green, London, pp 115–116
175. Mackenzie W (1830) A practical treatise on the diseases of the eye. Longman, Rees, Orme, Brown, & Green, London, p 234
176. (1845) A Descriptive and Illustrated Catalogue of the Calculi and Other Animal Concretions Contained in the Museum of the Royal College of Surgeons in London [Part II Calculi from the Urinary Organs of the Lower Animals and Part III Concretions found in Other Parts of the Body. RCS-MUS/7/19/4. Viewed at the Royal College of Surgeons, London, p 156
177. (1845) A Descriptive and Illustrated Catalogue of the Calculi and Other Animal Concretions Contained in the Museum of the Royal College of Surgeons in London [Part II Calculi from the Urinary Organs of the Lower Animals and Part III Concretions found in Other Parts of the Body. RCS-MUS/7/19/4. Viewed at the Royal College of Surgeons, London, p 264

178. Dear P (2006) The intelligibility of nature: how science makes the world. University of Chicago Press, Chicago
179. Lewis AJ (2011) A democracy of facts: natural history in the early Republic. University Pennsylvania Press, Philadelphia
180. Topham JR (2016) The scientific, the literary and the popular: commerce and the reimagining of the scientific journal in Britain, 1813–1825. NOREAY 20:305–324

8 The Meaning of Bodily Concretions, History, and Concluding Remarks

Bodily concretions can be formed in many different parts of the body, from the urinary organs to the lungs. Medieval surgeons and physicians proposed that there was even a stone that caused madness that formed in the brain [1]. The healing professions of surgery and medicine as well as chemists interested in understanding the life processes in the body felt that they had little to no good answers about how to prevent or better treat bodily concretions and help patients in tremendous pain. Physicians wanted to save their patients from the knife of the surgeon in the painful lithotomy procedure, while surgeons wanted to shrink stones so that they could improve a surgery that they often built their reputations. Chemists wanted to continue the wave of optimism of the Chemical Revolution that chemistry could solve real world problems and improve people's lives. Urinary stones were not an abstract problem like phlogiston or understanding the principle elements of life, but decomposing substances into their base parts and determining if there were substances that could remove these base parts from the body or prevent them from getting worse in the body. Chemists, along with physicians and surgeons, wanted to make a comprehensive collection of stones so that they could build full organizational schemes and improve their knowledge of their chemical compositions. The Enlightenment impulse of collecting was alive with bodily stones, much like planting or fossil collecting [2–4]. Chemists continued to describe and analyze these stones so that they could organize knowledge as perfectly possible, and many physicians and surgeons thought of this project as a major service for mankind.

Bodily stones were boundary objects, as they brought together seemingly discreet disciplines because of a shared focus on improving knowledge on understanding why the body produced calculi. Unlocking the secrets of concretions built up the reputations of physicians as scientifically minded, especially in regards to the application of chemical analysis to their bedside practice. Surgeons also benefited from working with stones, as they further validated their usefulness as providing the only convincing treatment for stones, but surgeons then turned to

E. A. Driggers, *Early Nineteenth Century Chemistry and the Analysis of Urinary Stones*, Perspectives on the History of Chemistry,
https://doi.org/10.1007/978-3-031-34973-7_8

contribute, as well as appear, in the medical and chemical literature through professional and informal interactions with physicians and chemists. Surgeons and physicians utilized stones as objects of debate in their intellectual societies and publications, in the hopes of spreading knowledge to their peers and interested parties outside the walls of the societies. Chemists benefited immensely from the discussion of stones, as they made themselves indispensable to societies that had collections that had become too expansive and unorganized. Practitioners often needed chemical instruction in order to be able to provide bedside diagnosis of the stones their patients were suffering from and consumed instruction books from medico-chemists. It would be difficult to think of a medical school today that does not provide chemical instruction to its students [5]. But surgeons, physicians, and chemists believed in limits to how effective chemical medicines could be to treating concretions.

The caution of chemistry continued through the nineteenth century, as practitioners did not want to harm the patient with uncareful use of medications. There was optimism that a substance might be procured in the future to provide immediate relief to sufferers of the stones, and acids were known to be helpful, but their consumption was dangerous. Chemistry was thought to be an effective tool in adjusting the fluids of the body. The concept of humoral theory was very much enshrined in the study of the chemistry of life and medical theory at the turn of the nineteenth century [6]. Humoral theory haunted the minds of chemists, surgeons, and physicians, even if they did not associate their ideas of the body's humors with the ancient ideas of ancient medical authorities like Hippocrates [7]. The physician was trying to end disease processes, like irritations, that produced stones by ensuring that the fluids could flow freely and were in a sort of balance. Surgeons, physicians, and even chemists used vocabulary and ideas from humoral theory in its nineteenth century version: humoral pathology. Ideas for treating stones in humoral pathology often involved the concepts of slow-moving fluids or the fluids themselves conveying the irritants that resulted in stones in the body. Humoral pathology continued to change over the nineteenth century, but the principle of balance and the importance of diseases originating in the fluids of the body remained [8]. Chemistry provided insights into what was going on in the body, as it was believed that the fluids could be read in the same way that physicians are able to see into the body today with a CT scan or X-ray. Urine, blood, and other bodily fluids could be studied and deeply analyzed with the tools and methods from the Chemical Revolution [9]. Information produced by chemical analysis of the urine could direct treatment or inform the providers that surgery for the removal of the stone was imminent.

In order for substances to enter the body and treat stones, physicians had to explain how medicines worked in the body. The idea was articulated by physicians like Edward Darrell Smith that substances changed very little when taken into the body. This philosophy of digestion was an artifact of humoral theory that was predominate until more sophisticated theories of digestion, aided by chemistry, were proposed in the in the 1830s [10, 11]. Chemical research seemed to confirm humoral theory and the idea that medicines and substances did not change in the body, as chemists were able to detect substances in the urine.

Neither chemists, nor physicians, nor surgeons had domination over knowledge about stones, and all three groups really felt the need to work together to improve knowledge. There were too many different types of stones and the chemical grouping of stones continued to change, which required all three groups working together. Surgeons were important on providing stones to physicians to study and investigate the history of the stones so that they could produce information about the number and population of patients that suffered from stones. Chemists had an opportunity to organize and train physicians and surgeons to analyze stones and fluids of the body at the bedside. No one group was able to control access to knowledge, as all three groups participated in intellectual societies and participated in those society's publication. Perhaps, the freedom of collaboration and cooperation was inspired by a medical marketplace that was loosely regulated and controlled, although societies like the Medico-Chirurgical Society and the College of Surgeons received royal patronage at the turn of the nineteenth century. There were many routs to also obtain chemical knowledge. Intellectual societies opened its doors to chemists, physicians, surgeons, planters, philosophers, but the membership was exclusively male and white, and mostly upper-class elites on both sides of the Atlantic. Some of these intellectual societies supported white supremacy and slavery as well.

Surgeons interested in chemistry changed the notion that they were practitioners only interested in craft knowledge. Many surgeons were interested in chemistry and embraced acquiring more knowledge, such as represented in the life of Astley Cooper and his friendship with Alexander Marcet. Cooper, and his surgical peers, wanted to continue to improve the lithotomy and make it less dangerous, as statistical studies had revealed it was a hazardous operation, but it still remained the only proven relief for urinary stones. Improved chemical knowledge was consumed by surgeons in the hopes of shrinking stones in patients. Surgeons needed extensive collections of bodily stones to train aspiring surgeons. Collections were also built to contribute to knowledge, as surgeons often kept personal collections of stones that they often shared with medico-chemists. The Royal College of Surgeons fulfilled that informal duty by building a comprehensive collection of stones that they wanted to share with physicians and chemists. Surgeons wanted to collaborate with other providers to build their intellectual reputations. The early collections of the College provided by John Hunter grew to include bodily concretions, as well as dinosaur bones. Participating in the material culture of science provided further professional and intellectual legitimacy to surgeons, who had previously been seen as craftsmen and beneath their university-trained medical peers [12, 13]. The College became an intellectual space of debate for many different professions, which further bolstered the reputation of surgeons, as spaces for discourse became an important part of science since the Royal Society [14]. The culmination of much of the chemical analysis and collecting efforts can be found in publications like *A Descriptive and Illustrated Catalogue*, which showed off the extensive collection of the College and highlighted the importance of chemists in organizing the collection. The College traded on the cultural capital of this collection and continued to show off their extensive collection of human bodily specimens, animal concretions, and animals for the training of students and the benefit of the public [15–17]. Museums cemented the intellectual contributions of a profession.

Physicians were interested in obtaining better chemical knowledge because they wanted to contribute to the healing of their patients and not merely pass them off to the surgeon. The Chemical Revolution revealed another tool for physicians to include in their repertoire and increase their value in the marketplace. The scientifically minded physicians advocated for the importance of chemistry, among other science, until the age of professionalization and marketplace dominance [18]. The tools the Chemical Revolution provided doctors embolden some physicians to denigrate other theories of medicine and promote what they considered the ideas of humoral medicine. Physicians began publishing chemical studies of their own, and other patient's urine, in the hopes of revealing the signs and signals for an imminent stone in the patient. They gathered stones and tried to organize them to explain disease and advocate for the danger of allowing patients to have internal fluid buildups. Important in understanding stones, was building collections that they could test, and physicians like John Yelloly and Alexander Marcet used personal contact to build extensive collections and collected figures of stone sufferers. Understanding the dangerous of lithotomies, who suffered from stones, and the number of admits to hospitals was important statistical information to collect and was helpful in underscoring the need to avoid lithotomies and search for chemical remedies that treated stone suffers [19, 20]. Information about patients, chemistry, and bodily stones was transmitted through cases that were published in medical journals or discussed in intellectual societies. Doctors who could not be present at society discussion could read of the latest ideas about stones, or even the promotion of humoral pathology, through journals and publications, that often included visual representations in artwork in the publications regarding stones. Some physicians embraced chemistry full and left the practice of medicine to concentrate on what they considered and important tool for healing and understanding the world.

The continued chemical research into bodily stones continued to validate and solidify the usefulness of chemistry. Chemistry was no longer the secret, clandestine discipline that was associated with alchemy [21]. The project of studying the chemistry of life, which included studying urine and concretions, had been part of the Chemical Revolution since the time of Lavoisier [22, 23]. Though chemists worked for surgeons and physicians, they were expanding their areas of expertise in chemically researching the stones. Medical chemistry is an important today in medical services in both America and the United Kingdom.

Humoral theory is caught up in the story of stones. Stones were thought to form through the irritations and problems of the internal fluid systems of the body. Understanding bodily concretions at the turn of the nineteenth century was caught up in trying to understand the humoral-medical mind at the turn of the nineteenth century. Physicians like Benjamin Rush, Edward Darrell Smith, and Thomas Cooper saw all disease in the body as explained by problems in these internal liquids. It is unclear what the nature of humoral pathology was at the turn of the nineteenth century, and it will take historians many years to understand the transitions in humoral theory that gets it to its state at the turn of the nineteenth century. Though the case of morbid concretions, or bodily stones, shows a snapshot of how physicians were trying to understand disease through the lens of fluid imbalances. In the nineteenth century mind, stones were a natural results of internal fluid problems that were either caused by external irritations that changed the fluids or diseases [24, 25]. The idea of sympathy showed that effects in one part of the body was reflected in another. Stress or diet could produce irritation in the body and then the fluids conveyed that irritation to the urinary organs, which might produce a stone. Rush thought of the body as entire connected and the fluids reflected changes in one part of the body to another [26], "The NEPHRITIC state of fever is often induced by calculi, but it frequently occurs in the gout, small-pox, and malignant states of fever." The fluids of the body continued to be researched intensely as there was enthusiasm that if the substances that caused stones could be definitively identified, outcomes for patients could be better as the appropriate medicines could be administered or surgery administered, but the emphasis on reading the fluids continued to be the forefront of research, which gave this research a character of humoral thought.

The ideas of humoral theory continued to remain in nineteenth century medicine. In the 1860s, William Thomas Brande, one of the chemists that was involved in analyzing the stones at the Royal College of Surgeons, published a *Dictionary of Science and Literature*, along with his co-author Rev. George W. Cox [27]. They included a definition of humoral pathology under the general discussion of pathology [27]:

> An important branch of pathology is that which treats diseases of the fluids of the body, and more especially of the disordered states of the blood and the urine (humoral pathology). This science has made rapid strides during the last half century, owing chiefly to the advance of animal chemistry, and to the application of the microscope in the examinations of diseased secretions and excretions.

The idea of the fluids being out of sync and in acrimony was an idea that was used to explain the formation of stones in the body at the turn of the nineteenth century. It was chemistry that provided an optimism, as well as other advanced tools like microscopes, that continued to make humoral pathology a viable way of studying the diseases of the body. Often it was the most advanced science and tools that continued to promote older ways of knowing, like humoral theory. But there is a lot of room for studying how the Chemical Revolution impacted the continuation of humoral theory. This book is one study of how parts of humoral theory continued to remain in medicine, chemistry, and surgery. This book also shows that even if a theory, as whole, falls out of favor, questions and insights from that same theory can remain in the study of diseases, or in some sub-communities. It was some of these community members shared interests in understanding stones that might have opened their minds to chemistry and humoral pathology, even if their interest in humoral theory only pertained to stones.

Anyone skeptical of humoral theory, or humoral pathology, simply has to have a physical, where the Complete Blood Count is run, and then the attending health-care provider tried to balance the excess and deficits found in the blood. During the creation of this project, I became ill, and my hemoglobin ran down to below three, unbeknownst to me, who felt fine. A CBC was run, and I was found to have a low-iron level, as I had problems absorbing iron because of my partial colectomy. My fluids were rebalanced with the administration of liquid iron and a blood transfusion. I felt amazing. The vital powers in my body were restored, and it was because of chemistry that my fluids could be read, and providers could intervene. The prospect of humoral theory often alarms people that I talk to about the major theme of this book, but they often want a practitioner of humoral medicine, wanting the doctor to ask [28] "how do you feel," then "where does it hurt?" The Kuhnian idea that though one theory becomes dominate while other theories still remain, though out of fashion, is an important one because it shows that in medicine, surgery, and chemistry, are only one challenging question, or perhaps interesting boundary objects, that could throw those unfashionable theories out of sub-communities, and back into the field of debate [29–32].

The most interesting part of writing this book has been being able to see like a historian, having an overview over the Atlantic that intellectuals and philosophers could study boundary objects like bodily concretions over time and space, and a rural practitioner who would be no more than a footnote in a parochial history of a frontier college could be a contributor to intellectual spaces thousands of miles away, and the problems of health are so universally interesting to other people, that his ideas could be picked up and discussed. In many ways, the life of Edward Darrell Smith represents the summation of this book. He was a curious person who trained at medicine, then traveled to study under Benjamin Rush in Philadelphia, then practiced medicine in rural South Carolina. He read the latest ideas of chemistry from Marcet and other Europeans and trained himself sufficiently to teach chemistry to college students, but continued to interact with intellectual societies, even across the Atlantic, and discuss their ideas in medical and scientific journals [33]. The desire to know and help suffers from urinary stones amplified ideas so that even in rural the Pendleton District, the broader ideas of chemistry, and even humoral theory, could be read and thought about.

References

1. Kao A (2000) Art of medicine: the cure of folly, or removing the "stone of madness." AMA J Ethics 2:22–23
2. Ogilvie BW (2006) The science of describing: natural history in renaissance Europe
3. Evans RJW, Marr A (eds) (2016) Curiosity and wonder: from the renaissance to the enlightenment. Routledge, New York
4. Plumb C (2010) Exotic animals in eighteenth-century Britain. PhD Dissertation, University of Manchester
5. Kohler RE (2008) From medical chemistry to biochemistry: the making of biomedical discipline. Cambridge University Press, Cambridge

6. Coley NG (2001) Early blood chemistry in Britain and France. Clin Chem 47:166–178
7. Lagay F (2002) History of medicine: the legacy of humoral medicine. AMA J Ethics 206–208
8. Ackerknecht EH (953) Broussais or a forgotten medical revolution. Bull Hist Med 27:320–343
9. Boantza VD (2013) Matter and method in the long chemical revolution: laws of another order. Routledge, New York
10. Alexa G (2010) Working ethics: William Beaumont, Alexis St. Martin, and medical research in antebellum America. Bull Hist Med 84:193–216
11. Mathias M, Moore AM (eds) (2018) Gut feeling and digestive health in nineteenth-century literature, history and culture. Palgrave Macmillan, New York
12. Blandy JP (2000) The royal college of surgeons of England: 200 years of history at the millennium. Royal College of Surgeons and Blackwell Science, London
13. Gelfand T (1980) Professionalizing modern medicine: Paris surgeons and medical science and institutions in the 18th century. Greenwood Press, Westport
14. Shapin S, Schaffer S (1985) Leviathan and the air-pump: Hobbes, Boyle, and the experimental life. Princeton University Press, Princeton
15. Findlen P (1996) Possessing nature: museums, collecting, and scientific culture in modern Italy. University of California Press, Berkley
16. Knoeff R, Zwijnenberg R (eds) (2015) The fate of anatomical collections. Routledge, New York
17. Berkowitz C (2015) Charles bell and the anatomy of reform. University of Chicago Press, Chicago
18. Warner JH (1997) Against the spirit of the system: the French impulse in nineteenth-century American medicine. Princeton University Press, Princeton
19. Frängsmyr T, Heilborn JL, Rider RE (eds) (1990) The quantifying spirit in the 18th century. University of California Press, Berkley
20. Porter TM (1986) The rise of statistical thinking, 1820–1900. Princeton University Press, Princeton
21. Principe LM (2015) The secrets of alchemy. University of Chicago Press, Chicago
22. Holmes FL (1987) Lavoisier and the chemistry of life. University of Wisconsin Press, Madison
23. Coley NG (2004) Medical chemists and the origins of clinical chemistry in Britain (circa 1750–1850). Clin Chem 50:961–972
24. Ishizuka H (2012) 'Firbre body:' the concept of fibre in eighteenth-century medicine, c. 1700–40. Med Hist 56:562–584
25. Driggers EA (2021) "The voice of the stomach": the mind, hypochdriasis and theories of dyspepsia in the nineteenth century. Hist Pysc 32:58–99
26. Rush B (1805) Medial inquires and observations, vol III. J. Conrad & Co., Philadelphia, p 58
27. Brande WT, Cox GW (eds) (1866) A dictionary of science, literature, & Art. Vol II. Longmans, Green, and Co., p 835
28. Foucault M (1975) The birth of the clinic: an archeology of medical perception (trans Smith AMS). Vintage Books, New York, p xviii
29. Star SL, Griesemer JR (1989) Institutional ecology, "translations" and boundary objects: amateurs and professionals in Berkeley's Museum of Vertebrate Zoology, 1907–39. Soc Stud Sci 19:387–420
30. Kuhn T (1962) The structure of scientific revolutions. University of Chicago, Chicago
31. Pickering A (1995) The mangle of practice: time, agency, and science. University of Chicago Press, Chicago
32. Pickstone JV (2000) Ways of knowing: a new history of science, technology, and medicine. Manchester University Press, Manchester
33. Driggers EA (2020) "A clue in the labyrinth:" the medico-chemistry of the stone in the life of Edward Darrell Smith. J Med Bio 28:232–245

Index

E. A. Driggers, *Early Nineteenth Century Chemistry and the Analysis of Urinary Stones*, Perspectives on the History of Chemistry,
https://doi.org/10.1007/978-3-031-34973-7